SASQUATCH DISCOVERED: THE BIOGRAPHY OF DR. JOHN BINDERNAGEL

Terrance N. James, PhD

SASQUATCH DISCOVERED

THE BIOGRAPHY OF DR. JOHN BINDERNAGEL

Terrance N. James, PhD

Copyright © 2022 Terrance N. James

Cataloguing data available from Library and Archives Canada
978-0-88839-751-5 [paperback]
978-0-88839-752-2 [epub]

All rights reserved. No part of this publication may be reproduced, stored in a retrieval system or transmitted, in any form or by any means, electronic, mechanical, audio, photocopying, recording, or otherwise (except for copying permitted by Sections 107 and 108 of the U.S. Copyright Law and except for book reviews for the public press), without the prior written permission of Hancock House Publishers. Permissions and licensing contribute to the book industry by helping to support writers and publishers through the purchase of authorized editions and excerpts. Please visit www.accesscopyright.ca.

Illustrations and photographs are copyrighted by the artist or the Publisher unless stated otherwise.

FRONT COVER PHOTOS: Chris Bindernagel & John Bindernagel Collection

PRODUCTION & DESIGN: J. Rade, M. Lamont

EDITOR: D. MARTENS

Crypto Editions is an imprint of Hancock House Publishers

We acknowledge the support of the Government of Canada through the Canada Book Fund and the Canada Council for the Arts, and of the Province of British Columbia through the British Columbia Arts Council and the Book Publishing Tax Credit.

Hancock House gratefully acknowledges the Halkomelem Speaking Peoples whose unceded, shared and asserted traditional territories our offices reside upon.

Published simultaneously in Canada and the United States by
HANCOCK HOUSE PUBLISHERS LTD.
19313 Zero Avenue, Surrey, B.C. Canada V3Z 9R9
#104-4550 Birch Bay-Lynden Rd, Blaine, WA, U.S.A. 98230-9436
(800) 938-1114 Fax (800) 983-2262
www.hancockhouse.com info@hancockhouse.com

DEDICATION

To the two Joans

CONTENTS

Acknowledgements 3
Introduction 5
Author's note 13
Part One: *The Man* 15
    Chapter 1 - *The young naturalist: "Bucolic times"* 17
    Chapter 2 - *Undergrad hopes: Wildlife conservation* 29
    Chapter 3 - *Uganda adventure: Game cropping* 37
    Chapter 4 - *Graduate work: Parasitology pinnacle* 45
    Chapter 5 - *Lost years: Candle and print making.* 53
    Chapter 6 - *Serengeti sojourn: Ideas entertained* 59
    Chapter 7 - *British Columbia: Domestic choices* 67
    Chapter 8 - *Iran: Persian politics* 73
    Chapter 9 - *Comox Valley: Home and abroad* 79
Part Two: *His Passion* 95
    Chapter 10 - *Serendipity: Evidence found* 97
    Chapter 11 - *Relevant scientists "They won't examine the evidence"* . . .109
    Chapter 12 - *Amateur investigators: "We owe them"* 149
    Chapter 13 - *Eyewitnesses: "They know what they've seen"* 185
    Chapter 14 - *Aboriginal community: "They get it"* 197
    Chapter 15 - *The media: "Uninformed or misinformed"* 211

Part Three: *The Legacy* 227
    Chapter 16 - *Toward the end: The last hurrah* 229
    Chapter 17 - *The Legacy: His footprint* 241

Appendices
    Appendix A: *Glossary*. 265
    Appendix B: *Dr. John Bindernagel Research Videos* 267
    Appendix C: *Filmography* 271
    Appendix D: *Index of Names* 273

Notes 279

About the Author. 293

ACKNOWLEDGEMENTS

This biography was written with John's input and with the support of his wife, Joan, and children, Chris and Sarah. John provided unusually rich sociocultural experiences for his family and they have shared details of family history, especially the international experiences with the United Nations Food and Agriculture Organization (UNFAO), which help us to know John as a family man and respected scientist. Their contribution to, and approval of this story, is greatly appreciated.

The encouragement and affirmation received from the sasquatch/Bigfoot community while researching John's story was heartwarming. While widely dispersed, this community shares a closeness bound by common experiences and beliefs. John was part of the community, and the community is part of his story.

Thanks is extended to those in John's email address book who responded to his final letter of January 8, 2018, indicating the imminence of his death. Your emails sustained him in his final days.

To John's friends and colleagues who responded to my request for input for this biography, and who have waited patiently to get their hands on this book, I offer my sincere thanks. Your photographs and stories enliven the text. Your love for John shone in your responses and I hope that it is conveyed in my writing.

Where permission was given, individuals have been named in the text. If anyone's name was left out it was because of a wish to be anonymous, or I erred in my record keeping. I carry the sole responsibility for the text and sincerely hope that I did not omit anyone who wished to be included.

Photographs in the text have come from a number of sources. Thank you to all who contributed. Please smile as you read your name in the text or attached to a photograph, and think of John. Any unacknowledged photos are from John's personal collection.

I am indebted to Alex Solunac for helping to get this project off the ground, Chris Murphy for orientation to the landscape of sasquatch research and editorial assistance, and Darryll Walsh for reading early chapters and providing encouragement. Special thanks is given to Robert Bateman and Boshkung Inc. for permission to include his "Sasquatch" painting which accompanied John's article in the Spring issue of *Beautiful British Columbia Magazine* in year 2000. Appreciation is extended to Andrew Benoit, Alex Witcombe, and Rictor Riolo, artists who have also allowed their art to be shared in this book, and to Gord Kurbis and CTV Vancouver Island for access to video files.

John always acknowledged the role of John Green as his mentor. Appreciation is expressed to Todd Prescott for providing access to, and permission to use, historical correspondence between John and John Green.

This book is dedicated to "the two Joans." John and I were each privileged to have a Joan as our help-mate. Without their love and support many things in our lives would not have been possible. In particular, I thank them both for their patience and understanding during the writing of this book.

I am grateful to Hancock House for adding this biography of John Bindernagel to their extensive collection of sasquatch titles.

T.N.J.
Courtenay, BC
April 2022

INTRODUCTION

November 10, 2017:

"I've been thinking about legacy," he said cocooned in cushions and a lap blanket.

"What do you have in mind?"

"A biography. My work isn't completed. The story of the unfolding sasquatch discovery claim needs to be put on record."

So we began to meet two or three times a week thereafter, at which times I fed him questions, recorded his discourses, and made notes.

But I am getting ahead of myself.

* * *

My wife and I were introduced to John and his wife, Joan, by our teen daughter, Karyn, a good friend of their daughter, Sarah. After a visit to the Bindernagel home one day, Karyn encouraged us to meet with them. The affirming "You will like them," was followed by an enticing "You should see the interesting animals Mr. Bindernagel keeps in the deep-freeze." This comment warranted further explanation. She described how she and Sarah went to the basement to get ice cream from the deep- freeze. When the lid was raised, a splayed bat mounted on a display board greeted her. Surprised, never having seen a bat before, she stifled a high-pitched exclamation. Sarah nonchalantly removed the bat to unbury the ice cream that had been placed there amid other frozen specimens.

John was working on his first book at the time: *North America's Great Ape: the Sasquatch*. I had just finished writing *Prader-Willi Syndrome: Home, School and Community* with Roy Brown. Immediately we shared a professional interest in writing. We were both researching and writing without university affiliation. Our disciplines were different, yet our research interests were very similar. He was listening to eyewitnesses tell of traumatic sasquatch encounters; I was hearing from families whose lives were dramatically changed with the birth of a child with a rare genetic syndrome. We were both recording stories and concerned with ecological assessments. My science was considered soft. I wrote from a psycho-socio-educational perspective; John's science was perhaps harder, although not considered "hard" in the rating of scientific disciplines. He wrote from the perspective of a very seasoned wildlife biologist. Our professional similarities far outweighed any differences. In fact, I don't recall ever disagreeing with John on anything. We shared a mutual respect for each other and our work.

For married readers, you will know that it is not always easy to find another couple with whom to relate where husbands and wives both enjoy a quality friendship. We were exceedingly fortunate in this regard. Our wives enjoyed many similar interests: gardens, birds, church life, reading, travel, grandchildren. They were also the same height and shared the same name—they were the two Joans.

As couples, we got together regularly for over two decades. In the early years we walked trails before stopping for tea or hot chocolate. Of late there was less walking and more time for ice cream sundaes. We had established an intimate level of trust as we shared our backgrounds. They had lived abroad in Africa, the Middle East, and the Caribbean. We had lived in Central America. John had worked with the Canadian International

Development Agency (CIDA) and the United Nations Food and Agriculture Organization (UNFAO). I had participated in a CIDA project in the Middle East and taught with the International Schools Organization in Costa Rica. We all enjoyed cross-cultural experiences. The two Joans maintained their multinational interests by volunteering with Ten Thousand Villages, the Mennonite nonprofit fair-trade organization that supports disadvantaged artisans in developing nations.

* * *

July 29, 2016:

It was a hot day as we went to McDonald's for ice cream. It was their favourite haunt, where they would often go to watch Canadian Football League (CFL) games on the large-screen television, because they didn't have a television at home.

This time, it was going to be different. My wife and I strategized beforehand. We already knew about John's cancer diagnosis and sensed that they might need to talk separately after their latest visit to the oncologist in Victoria. Often John regaled us with rather animated presentations on sasquatch activities and Joan had little opportunity to talk. We anticipated this day would have a different focus.

The most recent news: Stage IV inoperable bowel and kidney cancers.

Prognosis: six months without treatment.

We talked about a lot of things that day. John had been shocked by the Stage IV diagnosis but was optimistic about prolonging life. He had no fear of death; rather he had so much yet to do on the unfolding sasquatch discovery claim.

He professed to feel well. Yes, he would undergo chemo and radiation treatments, the schedule for which would require some adjustments to some commitments that he had already made, and a delay to the trip that they were planning to the Philippines. His speaking engagements would be cancelled.

John shared his concern for Joan. She had previously arranged their travels, researching the economic opportunities and taking care of the bookings. He had always been thankful for her executive management training and skills, but of late he'd had to take over all travel planning and financial management. Joan readily acknowledged that memory was giving her difficulty. He was already tying up loose ends and involving the children, Chris and Sarah, in their personal affairs.

Some people thought John to be the absent-minded professor type. To the contrary, his mind was an extensive databank, with instant recall. The problem was that his files were better organized in his head than in filing cabinets in their home. The house was a clutter of paper, with no logical system of retrieval for specific data, especially by others. The thought of organizing it all was now overwhelming. "I have to establish priorities," he said. The prognosis came with the harsh realities of impending decline in focus and ability. Importantly, he was in the midst of making a series of videos for YouTube which he needed to finish.

The topic of legacy was revisited. Over the years we had periodically addressed the hope that younger professionals would emerge to share our interests and carry on our work. There was no one on the horizon for John. The books and videos would have to suffice. There was a sadness—what to do with eyewitness reports still coming in. He wanted to support those who were so willing to share. He acknowledged his position of privilege and the trust that was given him by so many.

As always, he shared what he was working on. In addition to the videos, he was in the process of acquiring four sasquatch foot casts and photographic evidence which had been produced in the Sayward area of Vancouver Island in 1993. These excited him. And he was working on responses to a couple

INTRODUCTION

of Google alerts that he received on Bigfoot. These he felt obliged to do, in the interest of science. He felt that he "had the backs" of other scientists, the nonresponsive ones.

After years of a frugal lifestyle, he and Joan had made a couple of important quality-of-life decisions. They had a natural gas furnace installed to replace their dependency on wood. He no longer would need to buck, chop and store wood for the winter. And they bought a television so that they could enjoy the CFL games and other programs at home.

We talked about letting people know what was happening to him health-wise. He had never wanted health issues to dominate any conversation, something he felt happened too frequently when seniors got together. In the grand scheme of things, personal health was a low-priority topic. He preferred to talk about sasquatches, science, philosophy, religion, sasquatches, travel, food, books, politics, sasquatches. He didn't want his health concerns to interfere with his work, and he didn't want people to back off or lose focus because he was terminally ill. Maybe later, but not yet.

I was motivated to make post-meeting notes that day, something I had never done before with John. I didn't think about why. Taking notes was a professional routine, and I was on autopilot.

* * *

Over the next year John and Joan went to Victoria numerous times for his treatments at the Cancer Clinic at Royal Jubilee Hospital. Treatment progressed as could be expected, and while he didn't experience serious side effects, this very active man was slowing down. Yet he still carried his laptop and files to keep himself occupied during the week-long treatment periods. There was always another email to respond to or a draft to compose.

John was very self-conscious. As he lost his hair, he wore hats—at home when greeting guests, when visiting the homes of others, in his videos, and definitely in public. His early etiquette training dictated removal of headgear inside a building, particularly in someone's house. He struggled to hide what was happening to him.

Our get-togethers continued, but with less regularity. He wasn't leaving the house as much.

There wasn't the same robust enthusiasm that had been characteristic of our earlier times together. The welcoming and departing handshakes, however, were always there, along with a reciprocating hug for my wife.

By October 2017, John shared that he had exhausted all treatment options and was going on pain management. The end was becoming more imminent. There was now an urgency in the timeline.

For Joan, it was a shock. She had expected him to get better. She thought the treatments were working and he would get back to normal. Chris had to step in to assist with programming his dad's days, to accompany him to the doctor, to dispense his meds, and bolster him when he needed to be picked up.

* * *

November 12 , 2017:

I began my data gathering with John in earnest. He had to take medication regularly to maintain a constant level of drug in his body in order to prevent recurrence of pain. Gradually, he needed additional doses for pain that occurred between the scheduled doses. His work no longer drove his day; it existed at the mercy of medication—pain-numbing, mind-altering multicoloured pills.

INTRODUCTION

I would go with a set of questions, my recorder, extra batteries, clipboard and paper. Our sessions were held in his sunroom at first, a room overcrowded with everything sasquatch, memorabilia from their travels, and the clutter of everyday living. John was seated in a well-used, secondhand armchair and bundled for warmth. Joan, always hospitable, made sure we had a glass of juice. I was thankful that my time there gave her a break when she could feel free to leave the house. John could last about two hours at first. Tolerance and schedules were of paramount importance.

He began the first session by offering an agenda, all of which focused on his desire to tell the sasquatch story. As I refocused on my questions, he sensed where I wanted to go with his story. "Good, it's your book," he said. I felt this was both a sense of relief for him, as he didn't have to prepare, and affirmation for me, that he approved the direction the book would take.

I had been hearing the story of the sasquatch discovery claim intimately for a decade. I needed to ask questions about his personal life. John was a larger-than-life personality, and there had to be so much more to his story. I wanted to know about his childhood, his love of science, his education, meeting Joan, overseas assignments and more. I wanted to know the things that offered insight into his character and established his credibility as an internationally recognized wildlife biologist. If this was to be a biography, I had to know the whole story.

AUTHOR'S NOTE

Contrary to popular practice, "sasquatch" is not capitalized unless, following the rules of grammar, it begins a sentence, or it is capitalized in a quotation. Its use is parallel to that of bear, cougar, wolf, or any hominoid. While some argue that sasquatch can be either singular or plural, as with deer, elk, moose, readers will find both "sasquatch" and "sasquatches" (favoured by John) used as the plural form in the text.

The term "sasquatch" pre-dates the term "Bigfoot" and is used in the Canadian context, as John argued, to be more accurate with its aboriginal etymological history. Bigfoot is used in American references and is usually capitalized. Both terms appear in the text and should be considered synonyms.

PART ONE

THE MAN

His hero:

"As I grew up I was fervently desirous of becoming acquainted with Nature."

JOHN JAMES AUDUBON (1808-1851)

CHAPTER 1

THE YOUNG NATURALIST:

"BUCOLIC TIMES"

From his childhood, John had an intense curiosity about the natural world. His was not the average curiosity of an elementary-aged student, nor the obsessive-compulsive or perseverant enquiry of a hyper child. It was an intellectual curiosity, an inquisitiveness beyond his years that asked more questions, craved greater knowledge, and sought deeper understanding. In particular, he experienced a connection with nature, the sights and sounds of which brought him joy and left him wanting more.

Family background

Born in Kitchener, Ontario, in 1941, John was the only child of Albert and Mona Bindernagel, first-generation Canadians of Dutch descent. They experienced the post-war economy, a time of industrial growth, economic boom, and material gain. However, his father, a house painter, struggled with self-employment and the seasonal demands of the construction industry. His mother was a homemaker, content to manage the home with economy and cleanliness. John described the family finances as "working-class adequate." Albert and Mona would be in the first generation to benefit from Old Age Security, the universal pension for Canadians, introduced in 1952.

His parents were committed to John, their only child, having opportunities. His father, in particular, wanted him to get an education and better his lot in life. The new national prosperity meant that postsecondary education, not even a dream for his father's generation, could be a reality for a good student. John was not pushed to do well in school, but he was expected to do his best.

Rural lifestyle

The family lived in a well-cared-for single-family brick dwelling on the outskirts of town.

One doesn't have to go far to be in touch with nature in the Kitchener-Waterloo area. It has southern Ontario's largest watershed. The Grand River, with its many streams, wetlands, and woodlands, provides year-round recreational opportunities. The deciduous forests, wet climate, and rural lifestyle nurtured John's interests in hunting and fishing.

By junior high school, John enjoyed hunting with his father, one of the few things that they shared in common. The pursuit of game was an acceptable working class activity. Hunting for food and sport was a regular Saturday-morning

CHAPTER 1: THE YOUNG NATURALIST

discipline. He had a .22-calibre rifle for gray squirrels and cottontail rabbits, and a 12-gauge shotgun for grouse and waterfowl. He particularly enjoyed wing shooting, matching his skills with a smaller target in flight.

ALBERT AND MONA BINDERNAGEL IN FRONT OF THEIR KITCHENER HOME

Even when emphysema curtailed Albert's ability to be in the bush, he would sit in the cab of his truck with a coffee, facilitating his son's weekly adventure. He valued being part of the hunting fraternity and would forgo his own comforts to see the family tradition continue. Albert was quiet by nature and not a teacher—learning from him was "mostly osmotic," John related later in life. His father taught him to hunt, but perhaps more importantly conveyed a reverence for nature, with its flora and fauna, seasons and cycles. There was an implicit concern for stewardship.

While Saturday morning was set aside for father and son to hunt, Sunday morning they went to church together as a family, where John served as an altar boy as a young teen.

JOHN IN SURPLICE WITH HIS PARENTS

    Sunday afternoon was reserved for an outdoor family activity: fishing. It was a traditional day of rest, where John and his father fished for chubs, shiners, and sunfish, all of which had little food or sport value, and his mother spent quiet time on a blanket or in the vehicle. Fishing was as important as hunting and took them to streams and lakes year-round.

    John enjoyed being outdoors anytime. Summers were usually warm and humid; winters were cold or very cold. Each season offered new

CHAPTER 1: THE YOUNG NATURALIST

opportunities. He delighted in hunting grouse and waterfowl in the fall. In winter, he liked to follow the tracks of animals in the snow. Spring and summer were opportune for birding, finding nests and eggs. Fishing was an activity that could be done across all seasons.

Extended family members and neighbours were also outdoorsmen. As a 12-year-old, John was keen to learn how to skin squirrels and rabbits from his mother's brother, Uncle Walter. As a young teen he enjoyed spending summer weeks on the farm of family friends, where he learned about farming, the operation of farm machinery, and how to hunt groundhogs. While there, he also shot pigeons hanging around the barns. He always looked forward to invitations from Uncle Walter or Scott Thompson, an older neighbour, to go fishing. As an only-child, he was verbally precocious and comfortable with adult conversation. He became a keen observer of all things natural and always had questions about wildlife. The tutelage he received from these significant adult men in his life encouraged his interest in natural history.

The young naturalist

While in junior high school, on the recommendation of his science teacher, Nick Carter, himself a lover of botany, John joined the Kitchener-Waterloo Field Naturalists as a Junior Naturalist. What was it that the teacher saw in young John that welcomed him into the seriousness of such dedicated adult company? John recalled this as a pivotal point of encouragement. The invitation to join was "recognition and affirmation" of his interests in topics of science.

Being a Junior Naturalist was the starting point for his lifelong interest in birds. They were the easiest animal to study—easy to identify, with a wealth of field guides available. At the club he was introduced to the life of John James Audubon, the American ornithologist, naturalist and bird painter, who became one of his heroes. Audubon's extensive, detailed documentation of American birds in their natural habitat resonated with him. There was something about the beauty of feathers and the magic of flight that caught his attention long before he understood the complexities of their ecological contribution. He eagerly attended Audubon wildlife film nights, which were public events sponsored by the club to accumulate funds to purchase property.

As a young teen, he would go to nearby woods to explore and identify birds. He read about, and built, tree swallow nests. One of the first nests was accepted for use by a pair of swallows. This was a meaningful early experience. "Their acceptance meant that I could have impact," he later said. Insects, amphibians, and mammals were of interest also, but more complex for a neophyte naturalist to study.

John began to collect bird eggs to study. There was always the desire for more specimens. This was his first experience with research and classification. On his belt he carried a field guide in a leather holster which he had designed, crafted and riveted. At age 16, he used the savings from his paper route to purchase binoculars, a significant seventy-dollar expense, but an essential tool for birders. John described his teen years as "perfect days watching birds, collecting insects and learning the name of whatever I had discovered." They were good memories—"bucolic times," he said.

The student

By his Grade 9 year, there was considerable affirmation of his naturalist interests. At one point he volunteered information on birds in his science class, and thereafter the teacher referred to him as "the scientist." In today's vernacular, John would probably have been considered a "nerd." He was preoccupied with natural history pursuits rather than typical teen social activities.

CHAPTER 1: THE YOUNG NATURALIST

While others played on school and community teams and socialized, he put his energies into the pursuit of knowledge. He did not have many peer friendships. He valued independence and privacy, perhaps a result of being raised as an only-child. To the uninformed, he was simply "different"—a bird-watcher. Little did they know of the seriousness of his commitment to birding, a discipline that offered training in the morphology, shape, patterns, vocalizations, and technical care of birds. These were skills that would later be transferable to his broader wildlife research interests.

John described himself as an "adequate" student, "average in every way." Apart from science, his grades were mediocre. His interest in reading developed slowly—"late," by his own admission. He never did acquire an interest in fiction when in school; rather, his curiosity and excitement about learning led him to non-fiction reference material—only serious stuff. His mother was a reader of romance novels, and his father of westerns, while their son began to devour encyclopedic knowledge in his later elementary school years. He was a fan of *Mad Magazine* in his youth, the satirical magazine from the U.S. that lampooned popular culture and public figures. This was heady humour for a young teen.

As the number of specimens grew at home, John created a museum in the basement where he displayed all of his nests and bird eggs. He engaged in self-directed learning. Without training, he was already developing good curatorial practices. Family and friends knew of his interests and sometimes contributed to his collection. One day, a schoolmate arrived with the wing of a Great Blue Heron. He had shot the bird and viewed the wing as a trophy worthy of display. While they were not on the same page with wildlife protection, John did accept the wing as intended for his museum, but thereafter wrestled with uncertainties of guilt.

At times, peers from high school would come by with science questions. John was gaining recognition as a fount of knowledge on natural history. If he didn't know the answer, he did some research and provided feedback.

He was always driven to learn more and by a desire to please others. He enjoyed receiving this attention, but it didn't translate into close friendships or generalize to many social invitations.

Parent support

The basement became John's bedroom, close to his museum and terrarium, where he could enjoy the companionship of his frogs. His mother, who obsessively maintained a clean and orderly house, at first was ambivalent about her son's activities. She was tolerant if the mess was confined to one room. She reached a point, however, where she tried to discourage John's hobby, asserting emphatically, "There are enough bird nests and bird eggs in this house." According to John, the underlying message was, "This isn't what normal boys do." He understood that "she just wanted cleanliness, conformity and normalcy in her home, consistent with the Kitchener-Germanic culture." Despite what he perceived as a sense of disapproval at times, it was his mother who gave him his first field guide for birds.

Albert encouraged his son's education. "I don't want you to be a painter like me," he said to John. As a tradesman, he wrestled with the need for self-promotion, the long hours, and the cost of benefits. He was not pleased when his neighbour, a government bureaucrat with health coverage and a pension plan, encouraged John to consider the trades, to follow in his father's footsteps for a career. Albert hoped for something more for his son and knew that postsecondary education held the key. John was not receptive to the neighbour's suggestion, either. He was ambitious to live up to his father's expectations.

For Albert, summer was the season for exterior house painting, which precluded family holidays. For John, it was an opportunity to continue to explore. He roamed fields and woods—climbing, hiking, swimming, birding and fishing. He took black-and-white photos of his catches and wrote details on the back—e.g., this one, dated July 1958: "13 lb carp caught on deworm in the Grand River near Freeport on spinning tackle and on 8 lb test mono-filament line." As a teen he was already making detailed field notes.

When he obtained his driver's license at age 16 he went on road trips to Michigan with an aunt and uncle. His uncle was confident in John's abilities and slept while John drove. The new scenery was interesting, "but they had no interest in birds," recalled John. "They never wanted to stop and view wildlife."

CONFIDENT TO RIDE WITHOUT A SADDLE

Year round, when he had the opportunity to visit the farm of family friends, he enjoyed the opportunity to ride horses, even without a saddle. He was confident with horses, as he was with all animals.

ON STAGE AT EASTWOOD COLLEGIATE

Being from a working-class family, John was expected to learn to work at an early age. With parental encouragement, he had a paper route for the *Kitchener-Waterloo Record* when he was in upper elementary school. He waited for the papers at the paper shack with the other boys, but his route was different. While others rode their bikes and threw the folded newspapers onto front porches, he walked the halls of St. Mary's Hospital selling his papers.

In high school, he progressed to a part-time job packing and delivering groceries for a nearby store. Delivery with the company GMC panel van was every packer's goal, one which he achieved. John could handle responsibility and knew the value of a dollar earned.

Albert and Mona supported their son's interest in music. The first mass-produced electric bass-guitars appeared in the early 1950s and grabbed John's attention. During his teen years he learned to play the electric guitar, gained confidence, and even performed before the school. The accompanying photo was taken on March 5, 1958, on the stage of the Eastwood Collegiate auditorium in Kitchener during an assembly in which he played lead guitar for "The Lonely One" and "Detour," hit songs of the year recorded by Duane Eddy. Eastwood Collegiate is a public high school in East Kitchener that had been established in 1956.

University bound

John barely made the 65 percent average on subjects necessary for university entrance. While wanting his son to get more education, Albert was reluctant to let him apply for a bursary. He was a proud man, and a bursary application would reflect on his inability to provide for his son. His wife, the pragmatist who had never encouraged education, convinced him to support the application when John's path was chosen and it finally became necessary.

Mona had always hoped that her son would follow in the footsteps of other members of her family and become a Lutheran minister. She even sent John to visit his older cousin, James, himself a Lutheran seminarian, in the hope that he could persuade him to follow suit. Instead, John was encouraged to pursue his dreams. The decline of the whooping crane population was receiving public attention at this time, and this provided a focus for what John wanted to do in his future. Saving birds became his goal. For this, he knew he needed to go to university.

The school guidance counsellor challenged John to think of career possibilities beyond ornithology, a field he viewed as too narrow for realistic employment opportunities. By graduation in 1960, John was steered toward wildlife management, with the prospects of becoming a game warden or forester. Although unsure of the appeal of these specific jobs, the broader field of wildlife biology became highly attractive. John sensed emerging possibilities for working with animals and their habitat.

After graduation he registered at the Ontario Agricultural College (OAC). The OAC, the Macdonald Institute ("the premier home economics school in North America"), and the Ontario Veterinary College were in the process of amalgamating to become the University of Guelph. In the interim, the OAC became the Agricultural Department of the University of Toronto. The campus of the OAC was convenient, only 20 miles from home.

CHAPTER 2

UNDERGRAD HOPES:
WILDLIFE CONSERVATION

There was nothing at university that paralleled or replaced the early association with the Naturalist Club of John's high school years. While he enjoyed hunting and fishing with classmates on occasion, these were ad hoc recreational activities when time permitted. And there wasn't much time, as the academic schedule dictated daily life. However, he always found time to read, particularly The *Globe and Mail*, to stay abreast of all levels of news. He followed with interest the work of the Audubon Society. The year he entered university, they began documenting the decline of the Bald Eagle population, attributing this to the use of the insecticide DDT. Annually, he found time to participate in the Audubon Christmas Bird Count, a tradition in Canada since the inaugural count of 1900.

Wing shooting

To hunt was more than just a family tradition—it was a connection with nature, the flora and fauna of God's creation. It was more than just a recreational pursuit; there was a spiritual dimension, a subtle connection between understanding himself and his environment. There was an intimacy with the Creator among the trees, fields, lakes, and streams.

So how did he rationalize his enjoyment of wing shooting with the conservation of an avid birder? In class, he was taught to collect zoological specimens for museums. He enjoyed the collecting and archiving activities. He acknowledged a conflict between conservation and killing but placed a higher value on the collection of specimens. Being a long-time hunter, he had less difficulty with this than some of his classmates.

Summer jobs

A bursary alleviated some of the financial concerns related to university life. John lived at home for the first two years, then a year in residence on campus, and the final year boarding in the community. As for most students, summer provided the opportunity to fatten his bank account for the winter. The first year he took a job motivated by pay—the installation of concrete road curbs and gutters for a municipal contractor. In his second year, he landed a supervisory role for a new conservation project that involved designing trail systems and developing interpretive signs and brochures for the F.W.R. Dixon Natural History Area, close to Galt. This was the wildlife sanctuary purchased by the Kitchener-Waterloo Field Naturalist group to which he had belonged in high school. Its ownership and maintenance had been taken over by the Grand River Conservation Authority. It was an enjoyable outdoor summer, which encouraged him to believe he had made the right program choices. Here he was energized as he transitioned toward a career in conservation.

At the end of his third year, he was required to participate in a work experience term, which took him to Fort Frances in northwestern Ontario, under the supervision of a biologist with the Ontario Department of Lands and Forests. Fort Frances was a small community in the Rainy River District, across from International Falls, Michigan. Fishing was one of the major attractions to the area. John's project involved working with Conservation Officers to catch, tag, measure and release white sturgeon.

While at Fort Frances he spent a good deal of his time on the water, where he learned how to handle small water craft. This was his first real experience living on his own, where he was responsible for accommodation, food and managing new adult freedoms, including time with a girlfriend. He had ample time for sports recreation—fishing for pickerel, trout, and sturgeon, and exploring northwestern Ontario.

John described himself as "a typical university student on a summer job, lacking sensitivity to the way things were being done," which resulted in complaints to his supervisor by the Conservation Officers. His boss explained to him how he had to downgrade his enthusiasm. He got the message that he was "too keen" and was bogging the project down. There simply wasn't enough time to dissect fish, as he wanted to do. He had to let go of his personal research agenda, which was to record data on the otolith organ, looking for an age and size correlation. His self-directed research agenda did not fit with the work routines and schedules of his co-workers.

The ape-man experience

Apart from summer jobs, the real learning about wildlife was taking place in classrooms, where the canon of knowledge was imparted by professors with varying aptitudes and personalities. One experience in particular in the OAC wildlife program left an indelible memory. During his third year, in 1963, he went to the barbershop for a haircut while home for a weekend.

While waiting, he read an article by Ivan T. Sanderson in *True Magazine*[1] about a so-called "ape man" in British Columbia. Included was an artist's rendition of the creature. John found the article captivating. Sanderson was a Scottish biologist who wrote on nature, cryptozoology and paranormal subjects. Sanderson coined the term *cryptozoology* in the late 1940s to refer to "the science of hidden animals."[2] Most definitions include the search for, and study of, animals whose existence is unsubstantiated, for example Ogopogo, the Loch Ness monster, yeti and sasquatch.

John attempted to raise discussion of this unknown species, the ape-man, in class, but the professor smirked and chuckled while his classmates outright laughed. Thereafter, he could not forget this magazine article, nor the classroom experience.

"There I was, a 22-year-old, an idealistic student of science, thinking that science is very open-minded, open to new ideas such as this. That I felt disappointment was to put it mildly," admitted John. He was surprised by his classmates. Was it peer pressure or obeisance to the professor? Were they not curious? Did they not value the inquiry of science? That day he learned about the game of science monopoly, where to take a risk might not allow one to advance, or go on to success.

Intrigued by the article he had read, he started to do as much research as he could on campus with magazines and journals to learn more about the ape-man, but there was very little to be found. He was disappointed again. The article had described William Roe's 1955 account of having a chance to shoot at the creature but not being able to pull the trigger. As an avid hunter, John couldn't get the story out of his mind. To have an animal in the cross-hairs, a sure shot at the prey, is a hunter's delight. To not be able to pull the trigger was almost incomprehensible. It was the seed of doubt that grabbed him. Roe said it was not a bear, but rather a human-like female creature. But what was it? Where would this creature fit in the taxonomy of mammals? Was this a new discovery? Was it related to the gigantopithicus?

Was this a great ape in North America? Such questions remained in the back of his mind thereafter.

An important mentor

Not all professors were as narrow-minded as the one who prevented any discussion of the ape-man. Dr. Anton de Vos, a wildlife biologist, taught about birds and mammals of North America with competence, but more important, inspired his students with stories from elsewhere in the world. He had worked and travelled in Africa and Dutch colonies and brought a broader world view to the classroom. He planted the seeds of possibility, of working overseas in wildlife conservation and management. There was a romance in his presentations: the excitement of mystery and discovery, and remoteness from everyday life, that captivated John.

GRADUATION – 1964

Until his fourth year, John anticipated a domestic career in wildlife management. He was being trained as an applied biologist, to take the results of academic research and apply them in the field. He was disappointed, however, with an interview with the District Wildlife Biologist to explore the field of wildlife conservation. He learned that the government focus was on the harvesting of wildlife, either for meat or trophy, as a means of wildlife management. "Our profession is not here to protect wildlife, we are here to facilitate harvest," the biologist stated. This was a controversial approach in the court of public opinion, and one which chafed at John's own view of what was needed. "The prospects of working on bag limits was distasteful," he said later. The implication was clear, however, that if he didn't like the focus he needed to look elsewhere for a job.

In 1964 John received a B.S.A. in Wildlife Management from the University of Toronto, as the newly formed University of Guelph was not yet granting degrees. Things were beginning to happen in Canada about this time—environmentalism and conservation were in their infancy. Rachel Carson's *Silent Spring*, an indictment of the careless use of pesticides south of the border, had been published two years earlier. The impact on the natural environment of air and water pollution, and hazardous wastes was raising alarms in Canada. Conservation, previously only championed by naturalist groups, was becoming a common concern. In 1962, the Canadian Wildlife Federation was born when conservationists from seven provinces met in Montreal. The next year, the group was urging government investigation of the effect of biocides on wildlife and pushing for restrictions on their sales. In the same year, the National and Provincial Parks Association of Canada was established. There was promise for new grads. However, Professor de Vos, John's respected mentor, let him know about a job available in Uganda as a wildlife adviser with the Canadian International Development Agency (CIDA). John had enjoyed de Vos's classes, the vicarious romance of travel abroad, and the first-hand stories of foreign birding. He applied, used his mentor as a reference, went through the interview process, and was hired. Without this first opportunity, John might have taken a different career path.

Parent anguish

The previous year, his father had died. John was living at home at the time and expected him to be coming home from the hospital, but he didn't. Prostate cancer took his life at age 73. John described his father's death as coming "out of the blue." He was unprepared for it. He had failed to understand the seriousness of his father's condition.

His mother, ten years younger than her husband, was widowed with only one child, and now he was going to Africa. There were extended-family members who thought he shouldn't be going. He was faced with a dilemma, his first experience with cognitive dissonance: should he stay home or go to Africa? What did an only-child owe to his widowed mother? What did he owe to himself and his future?

John went, believing that his mother had a strong support network of family and friends. She returned to work as a clerk in a grocery store and wrote many letters to him in Uganda. In later life, he admitted that his mother probably never got the acknowledgement she deserved. "I would bring university classmates home after a morning of hunting, and she would provide a generous lunch," he said. Her hospitality was an obvious source of pride for him. He also praised her housekeepng and care of her family.

While John was at university, he was always nearby. Even when living in residence, he still came home on occasional weekends and was there for family get-togethers. On the surface, it was a decision to go to Africa for just a year. Little did anyone realize the importance and implications of this step.

Meeting Joan

While the job prospect was exciting, there was a competing interest. John had attended a church youth activity, a hayride, with a chum and connected with Joan Keyes, an attractive young woman who worked as a librarian in the Apiculture Department at the same university, a place that he frequented. She was a graduate of the University of Western Ontario and had trained to be an executive assistant. She was always organized.

In reflecting back on that time, Joan said she found the things John was doing to be "interesting." She also had a bent toward science.

Her involvement with a department of science and openness to exploring and learning suggested a compatibility of interest and intellect that pleased John. The attraction was mutual, and they had begun dating. Now he had to tell her he was going to Africa for a year. This was not an easy decision for him. He saw the prospects of a long-term relationship with Joan. While telling his mother he was going to Africa was not easy, he knew that he could survive without her and others would take care of her if needed. Leaving Joan was different. There was greater risk and emotion.

Neither wanted it to end, so they agreed to a long-distance correspondence connection.

CHAPTER 3

UGANDA ADVENTURE:

GAME CROPPING

While the Uganda Independence Act, passed in 1962, gave Uganda independence from the United Kingdom, history shows that the year that really changed the country was 1964. A mutiny of Ugandan Army soldiers led to the rapid promotion of a minimally educated junior officer, Idi Amin. John arrived in Uganda in 1964. It was a country divided—along national, religious, and ethnic lines.

CIDA project

As a Canadian International Development Agency (CIDA) worker, John was responsible to the Chief Game Warden of Uganda, a man with many years of experience in the country. His position was a remnant from the British colonial period that had recently come to an end. John was to work with Ugandan game wardens to explore game cropping possibilities as a means of meat provision and wildlife conservation, instead of the introduction of cattle, which had already been widely established on suitable non-game areas in the country and was taking over areas outside of the national parks.[1]

Unfortunately, game cropping had the taint of poaching, an historical problem in the area. Community members who depended on wildlife for meat and ceremonial dress used to hunt both inside and outside of protected areas. Wildlife had no regard for boundaries and hunters followed them. Some game animals with high tourism and trophy value, such as lions and zebras, were problematic in their association with graft and corruption. The focus for game cropping was on lower-value animals, particularly wildebeest and African buffalo. The buffalo offered a large meat catch at one time and was more pleasing to the palate of the people than the wildebeest. Locals liked game meat but would not pay for it when it could be poached. Villagers did not understand the economic benefit of game cropping as the central government did.

The work undertaken in Uganda became the basis of John's master's thesis. Part of the fieldwork involved assessing the overall health of animals and collecting parasites from their hosts. Tapeworms attach to the intestinal wall and live on nutrients from digested food. The accompanying photo of John displays a 96-foot-long tapeworn taken from the small intestines of an African buffalo.

CHAPTER 3: UGANDA ADVENTURE

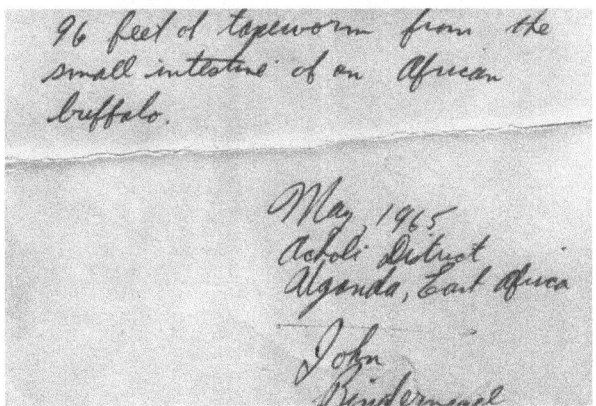

A house was provided in a compound that housed other international workers. The assignment was exciting and rewarding. John's hunting background helped him understand local hunting issues. He was delighted with the new animals he was learning about. But there was something missing. Daily he went back to the camaraderie at the compound, but to the loneliness of his house. Joan was on his mind.

Marriage

WEDDING DAY – NOVEMBER 27, 1965

It didn't take long before John proposed to Joan, by mail. She accepted with one condition: that she wouldn't send out wedding invitations until he came back to Canada on leave. When Joan announced her engagement to her parents, they were surprised and asked, "To whom?" Joan had been living in Guelph and her parents in Kitchener, so she had been seeing John without their knowledge. John came home on one month of leave, and wedding planning proceeded quickly. They were married on November 27, 1965, in the United Church that Joan had attended as a youth in Kitchener. "My parents thought the quick marriage was scandalous," Joan said with a laugh.

Joan's father had been an elementary school principal, and summer holidays took them as far as Detroit to visit friends. She had no international travel experience other than such short trips to the U.S. but was excited at the prospect of seeing Africa. The move, however, was hard on her parents. They had lost their son a decade earlier. He had been riding his bicycle at night and was hit by a car. Now it felt like they were losing their daughter, their only child.

A sense of adventure

The newlyweds returned to Uganda to live in the same house that John had already occupied, a colonial-style structure with a big yard and a little cabin at the back for a houseboy. "Houses were quite far apart," said Joan. "We hardly ever saw our neighbours." The houseboy cleaned the cement floors, did some shopping and cooking, cut the grass with a machete, and guarded the place when they were away.

John established a camp about two hours distant where they would stay for a week or so each month. There was a guard at the camp when they weren't there. From the camp, John would hunt and harvest wildlife with local men and make wildlife observations. The men would shoot from the Jeep, often with their rifles braced on the roll bar, with John still sitting in the driver's seat. He attributed his eventual hearing loss to cochlear damage from this close proximity to the gun blasts. John enjoyed time in the field. He is shown on the next page with a warthog. Joan managed the camp, enjoyed the wildlife and scenery, but had no interest in participating in the hunting.

JOHN WITH WARTHOG

Joan experienced a sense of adventure that was new to her. John had a motorbike, and she enjoyed the ease and freedom of travel as his passenger. "We had to go through the national park to get to the research station," she said, recounting how one day they suddenly came to a halt, face-to-face with a towering elephant on the narrow road. "The elephant wasn't moving," she said. "So, John moved the motorbike forward and back a few times, but still it didn't move. Then he revved the engine and the elephant raised its trunk and bellowed." The elephant controlled the road and was not about to move. "John told me to hang on, and he revved the engine again. We sped narrowly by. Life was exciting at times," she said, laughing.

They both recalled that it was a great way to start married life—the excitement of the job, the new travel explorations together, and the absence of influence from mothers or mothers-in-law. "It was like an extended honeymoon," John said with a smile.

Political climate

A couple of Canadians working in London came to visit them, and the two couples toured East Africa together. While on the road there was a coup in Uganda. "This was a tribal issue," said John. He did not feel at risk, although relatives back home assumed the worst from news broadcasts. A British member of the international team working on tsetse fly control was injured in a violent *panga* (machete) attack, but John never felt serious concern for their personal safety. Joan, on the other hand, admitted that she was afraid when they were in the street and people stared at them. Once, when they were in town, people mobbed their vehicle, and she didn't know how to react. "They were just curious about the unusual people in unusual clothing," John explained. But the reality was that civil unrest was increasing across the country.

Looking back, John admitted he underestimated other dangers while living in Uganda, and mentioned malaria, invader ants, and swims in rivers inhabited by crocodiles. Food was never refrigerated—"the reason for curry in the tropics," he explained. Here he acquired a taste for curried foods and tropical fruits, especially papayas and mangoes.

John asked for an extension of the original contract, and they stayed an additional two years, until 1967. The assignment came to a natural end at that point, and he was not replaced. The project question had been: Can a portion of the animal population be responsibly harvested?[1] By the time he left they were still trying to determine what was reasonable. Even among the international team, there was skepticism about the project. John felt satisfied with his day-to-day contribution, particularly the staff training that he had done, "but there was little impact on government policy," he said.

Heading home

It was hard to leave friends that they had made, including the cooks and houseboy, who had taught them a great deal about the culture. "They were always very patient with me," reflected John. They had enjoyed the rainy, tropical climate with its two dry seasons and an average annual temperature of 26 degrees Celsius (78.8 degrees Fahrenheit). The scenery, the central plateau, with its miles of lakeshore and rim of mountains, was always pleasant. The work among an international team of scientists was stimulating, and the cross-cultural experiences rewarding. This first overseas assignment was indelibly etched in their memory. As they flew out of Kampala, they were looking forward to returning home—and they were heading there with a new plan. However, they both knew they would be open to overseas work again in the future.

In February 1966, a year before they left Uganda, Milton Obote, the Prime Minister, suspended the constitution and assumed all government powers. He had promoted Idi Amin to Commander of the Army the previous year. In September 1967, the year that John and Joan left, a new constitution proclaimed Uganda a republic and gave the President even greater control. Obote was deposed in a coup at the beginning of 1971 by Idi Amin, and an estimated 300,000 Ugandans subsequently lost their lives under his dictatorship.

CHAPTER 4

GRADUATE WORK:

PARASITOLOGY PINNACLE

John knew that involvement in wildlife research, his newly-defined career goal that emerged from his Ugandan experience, would require an advanced degree, probably a PhD. When he was taking a veterinary course in parasitology, one of his last undergrad courses, he had heard about a zoology program of interest at the University of Wisconsin–Madison. His grades had improved during his undergraduate years and he chose to pursue graduate studies when he came home, with the prospect of becoming an expert in a narrow and manageable field of scientific interest.

University of Wisconsin–Madison

John applied, was shortlisted, and was one of four students selected for the graduate program in zoology. The fact that he had studied physics, a subject not required in the American system and unrelated to the field of study, "impressed them and spoke well of the Canadian education system," he recalled with a laugh.

During his three years at the University of Wisconsin, John completed the requirements for an MS in Zoology and a PhD in Zoology/Veterinary Science (1970). His doctoral thesis was titled: *Abomasal nematodes of sympatric wild ruminants in Uganda, East Africa*.[1] The Acholi Game Cropping Project that he had been involved with in Uganda, experimentally utilizing free-living ruminants as a source of meat, provided the framework for his study. The harvesting of large numbers of such animals in one area presented an unequalled opportunity to collect parasites from their hosts. This work was important "in view of the global food shortages and particularly of protein shortages in many African countries." In the introduction to his study, John pointed out "the importance of gastrointestinal helminths as a deterrent to efficient meat production in both domestic and free-living ruminants," and how they "must necessarily become a consideration as the more conspicuous protozoan, bacterial and viral diseases are brought under control." Since the abomasum, one of the most important organs in the ruminant digestive tract, was known to be parasitized by some relatively pathogenic nematodes in livestock, his study was directed at abomasal parasitism in wild ruminants.

He and Joan lived in student housing in Madison. The oldest and largest public university in the state, the school had a fine research emphasis. John's life was focused on the lab and classroom; Joan's was centred in their one-bedroom apartment, with its tiny kitchen, small living room and very limited bathroom. There were six couples in their married student housing block, with similar lifestyles and with whom they socialized. Typical of wherever they lived, they also made new friends at church. Joan helped John in the lab, kept him company there many evenings, and did typing for him. And now that they were closer to home, they enjoyed some parent visits.

In their final months the campus was rife with student activism—firebombings, protest marches, crowd rampages, and a teaching assistants' strike. At Madison the Army Mathematics Research Center was targeted by students as a protest against the university's connections with the U.S. military. Fire bombings resulted in the National Guard being called to campus and the use of teargas against protesters. The killing of students protesting the Vietnam War at Kent State University occurred a month before John's commencement ceremony and was the catalyst for protests and violence across many American campuses. Over $2.7 million in damages was done to at least six buildings on the University of Wisconsin–Madison campus.[2]

First child

Joan admits she was oblivious to what was going on politically around them. John was focused on his research and dissertation and she was supporting him. There wasn't much time for recreation, but on one occasion they were invited to a cabin on Georgian Bay owned by a classmate's parents. "We were in a power boat, bouncing across the water, when I noticed Joan looking queasy and holding her stomach," said John. "Innocently, I asked: 'Have you got something in there?'"

Her reply was a "complete surprise."

"Yes, a baby."

"We came home smiling like a couple of silly kids," John recalled. Recounting the story put a wide grin on his face again.

Their first child, a son, was born on March 23, 1970, at the university hospital during their last semester in Madison. They both liked the name Christopher, the name of a research colleague who had worked with them in Uganda.

Post-doctoral fellowship

As graduation came closer, John's expectation was to work within the field of veterinary parasitology. He was committed to studying the origin and development of parasitic infections in animal hosts. Given that parasites can carry diseases that might prove fatal to domestic and wild animals, and that they can also be transmitted to humans, there was the very satisfying secondary benefit of contributing to public health.

As he could no longer stay in the United States on a student visa, John began to explore job possibilities in Canada, including with the University of Guelph, now a full-fledged degree-granting institution. Here he was offered a one-year postdoctoral fellowship with Professor Roy Anderson, which he accepted. The project involved collecting and analyzing larvae of the meningeal worm from white-tailed deer scat. The fellowship combined his interest in parasitology with his longstanding interest in wildlife field work. The Canadian Wildlife Service contract to survey wildlife parasite distribution took him across Quebec, Ontario, Manitoba, Saskatchewan, and into Alberta. He found larvae indistinguishable from those of the meningeal worm as far west as Alberta, and there was no apparent barrier to their continued spread westward. The parasite was established in white-tailed deer of the aspen parkland. This was John's first real foray into mainstream science and the only parasitological research he ever undertook.

Joan and their infant son accompanied him as they camped while doing field work in the fall and winter of 1971. Once again, space was cramped and the weather was cold, but they were happy to be outdoors. Some analysis was done on site and some frozen scat was taken back to a lab in Saskatoon for storage and study. It was here that John met Don Blood, Chief Ecologist for Saskatchewan. "He used our lab in Saskatoon to store equipment and samples," Blood explained. This contact would factor into John's later story.

IN CAMP WITH BAERMANN APPARATUS

The accompanying photo shows John with a Baermann apparatus, consisting of a funnel containing muslin filters for straining out larvae from fecal specimens. The larvae wiggle out of the feces, cannot swim against gravity, and fall through the water to the area clamped off by tubing, which is released to collect the sample for identification.

Research and authorship

Two papers were published from this work in 1972, co-authored with R.C. Anderson, the project supervisor.[3,4] Whether it was a misunderstanding or simply naivete, hard feelings developed over a problem widespread in science: authorship. While the project and its design belonged to Anderson, the research and the writing were done by John. In science, there are no precise guidelines for authorship. Most understand that credit must be limited to those who have contributed substantially to the work; however, authors can contribute in various ways. To complicate the problem further, different disciplines have adopted their own practices for the order of naming authors. In short, was John's contribution to be senior authorship, co-authorship, or junior authorship? He did have one additional publication credit related to this study under his own name.[5]

Publication is a very important part of a professional role in science, particularly in academia.

"I was young and feeling territorial," John related, and this led to uncomfortable friction. "Is this what it is going to be like?" he questioned. "Do I want a life filled with this type of tension?" The issue represented by these two papers was part of the larger "publish or perish syndrome," which exerts pressure in academia to rapidly and continually publish new work to further one's career. Advancement up the academic ladder to obtain the highest rung of tenure depends on publications as a demonstration of academic talent.

Two additional papers were published in reputable scientific journals based on his work in Uganda.[6,7] John was gathering publication credits and establishing his professional credibility, but something didn't feel right.

Disillusionment with science

The project drew to a close by mutual consent and ended reasonably amicably. However, the experience of this postdoctoral project planted seeds of disillusionment about the professional role of a scientist. So much so, that John began to question his career choice. He did not particularly enjoy the restrictions of academia, the long hours of sterility in the lab, nor the politics of professional science. He realized he was not destined for an academic, bureaucratic or teaching career.

Expertise in parasitology, such a narrow area, was not fulfilling. As a scientist his heart was still in the fieldwork of an applied biologist, his undergraduate degree, where he always believed he could make a difference. It better encompassed his love of animals and concern for their conservation. But even that fire was quenched at this point. For the second time in his life, science failed to meet his expectations. He was feeling trapped. He had enjoyed the intellectual challenge and personal growth of his graduate and postgraduate work, but he was not in the right frame of mind to continue. He was feeling the negative effects of disappointment.

Since he first recognized his love of science, he had always had high expectations. He had put his heart and soul into postgraduate work, but his relationship with professional science was unrequited. His idealism thwarted happiness. For someone already a little higher on the scale of hyperactivity, this produced increasing anxiety. His heart was in the resource management and conservation of wildlife; his head was in the natural environment, not in a classroom, lab, or government office. Life was not going according to Plan A. Change was needed.

The summer was spent helping with a study of mountain caribou in Jasper National Park, an opportunity arranged by Blood, who was then with the Canadian Wildlife Service in Edmonton. David Thompson, a surveyor and trader for the Northwest Company, had in 1811 recorded unidentified animal tracks measuring 14 inches in length and 8 inches wide near Jasper.[8] It is uncertain if John knew about this report at the time.

    Chris's earliest memories of being in the field with his dad stem from this summer in Jasper: "My mom and I would come along on the field study expeditions, hiking or riding on the pack horses and watching for animal droppings."[7] John delighted in having his family with him in the field, sharing his experiences. However, the summer and the contract ended—and then there was Plan B.

CHAPTER 5

LOST YEARS:

CANDLE AND PRINT MAKING

The hippie movement of the 1960s, associated with facial hair, tie-dyed garments, and a back-to-the-earth mentality, was waning but still inviting. In the vernacular of the day, John "dropped out." He abruptly escaped the previously promising career of parasitology, and along with it a lifestyle dictated by mainstream science. It was a quiet rejection of the lock-step approach to career development, the climbing of the academic ladder of success. He was later to refer to the next two years as "lost years," a reference to the resulting gap in his curriculum vitae.

Candle and print making

He and Joan took up print making and candle making, seeking expression of their artistic sides. The candle, like a campfire, was a reflection of the past, a symbol of a simpler era when humans co-existed with nature rather than tried to control it. There was something earthy, traditional, and wholesome about engaging in candle making.

This creative activity was quiet and peaceful—and disassociated with the rigours of science. "John was the imaginative and creative one," reflected Joan. He began to explore "gyotaka," the traditional Japanese method of fish printing, and experimented with printing from wood and lino cuts. The burlap sandpiper print on the next page was made from salvaged cedar barnwood. It utilized square nails for the beak and legs. The "Masai" print is a lino-cut made several years later and illustrates John's longstanding interest in print making as an art form. Unlike candles, prints were decorations that had limited consumer appeal.

Candles at least had a utilitarian purpose. Candle making became a metaphor for their drastically different lifestyle. They went from the stimulation of science, with its established investigative processes and taxonomies of knowledge, to the simple serenity of dipping wicks into paraffin. However, "we were ill-equipped for the demands of the farm-market economy," recalled John. Their candles reflected a working-class utility rather than avant garde artistry. Their product wasn't particularly valued by the public, other than as emergency lighting during blackouts, when in reality any candle would do. Money became tight—there were supplies to buy and table space to rent in order to sell their products at markets and craft fairs.

John was a hippie at heart. He was "hip," more aware of the socioeconomic and geopolitical circumstances of the world, than most of his peers. He had experienced the realities of rural Africa, the impact of the Vietnam War in the U.S., and the attraction of university students to peace, love and non-materialism. Hippies opted for a simpler, more peaceful lifestyle, values

CHAPTER 5: LOST YEARS

to which he and Joan were attracted. He sported long hair and a beard and fit the countercultural image, but he didn't participate in the drug culture.

SAMPLES OF JOHN'S EXPERIENCES WITH PRINT MAKING

As the months went by, they experienced the economic marginalization common to hippies. It took them a while to realize that this was an uneconomic endeavour and that they were paying too much to indulge their new freedoms. While philosophically leaning to an alternative lifestyle, it became obvious that the hippie model was not going to work for them.

Despite the economic hardships, the change of routine was good for John. "It took me a little

while to recover from the negative aspects of doing science research in a university environment," he recalled.

At the same time, there was feedback from family, the type uttered behind one's back. "John has a PhD and isn't using it." The message was irksome, but did cause John to question what they were doing, and why. He came from a family culture where quitting was never the right thing to do. The old adage, "When the going gets tough, the tough get going," was indelibly etched in the back of his mind.

Time brought healing; the burnout abated, and a fresh optimism emerged. While he was greatly disappointed with the strictures of the ivory tower and professional roles of science, he hadn't lost his love of the natural world. It was time to look for work.

Overseas again?

The time spent in Uganda had represented a balance—the chance to learn and explore, while at the same time to contribute and make a difference. They had good memories of their first years abroad and began to think of overseas work again.

John had a former colleague living in England who he hoped might be able to put him on track to something. He went to England alone for a visit, with Joan's blessing, and then on speculation went to Rome to explore possibilities with the United Nations Food and Agriculture Organization (UNFAO). Professor de Vos, who had let him know of the previous CIDA opening in Uganda, was now working as a Senior Wildlife Advisor with the UNFAO in Rome.

"I had a strong interest in returning to East Africa to work in wildlife research, conservation, and management and thought I would look up my old mentor and try to find a similar assignment in the UN system," recalled John. On the strength of his previous CIDA experience in Uganda, "and a very informal interview," he was offered a wildlife advisor position with a project in Tanzania. They were going to the Serengeti, the vast 12,000-square-mile (31,080 square kilometres) plain in the north of the country, to the west of the Great Rift Valley. The Serengeti is a delight for wildlife biologists with its protected game reserves, conservation areas, and the largest mammal migrations in the world.

They were excited to be going back to Africa; however, they had to wait what seemed like an interminable length of time in a small, rented house on Bingham Street in Kitchener for the paperwork to arrive. The uncertainty of the timeline discouraged local employment and drained their meagre resources. When the contract eventually arrived, they said goodbye to family once again. This was a doubly-difficult thing to do, as both were only children. They were abandoning their parents once again.

John's contact in Rome served him well and kick-started his UNFAO career, which would take him to many countries in the future.

CHAPTER 6

SERENGETI SOJOURN:
IDEAS ENTERTAINED

The Serengeti Research Institute was within Serengeti National Park, about 325 kilometres from Arusha, the gateway to safari destinations in Tanzania. It was the forerunner of the Serengeti Wildlife Research Centre. Today the park is one of the largest wildlife sanctuaries in the world and a World Heritage Site and Biosphere Reserve.

The Serengeti Research Institute (SRI)

Life at the SRI was "fairly primitive," according to John. Accommodation was provided by the Tanzanian government in a compound with other international scientists and their families. One diesel generator was shared between two families. It took some negotiation with the neighbour to determine how late they could stay up at night, which was difficult at times. John preferred the benefit of light late at night so that he could stay up and write. When it was time to shut down, someone had to go and turn off the generator. This involved a 100-metre trek through bush where lions, hyenas and snakes were about. Flashlights with fresh batteries were always needed. Sometimes the neighbour wanted to close down earlier. There was always the question of whose turn it was, and what time was going to be "lights out." When they ran out of diesel fuel, it was dark by 7 p.m. Then it was flashlights and batteries, which were limited in supply and expensive. "One of my fondest memories of Africa was going out with Dad to fire up the generators. This had to be done to keep the place operating with electricity every evening," said Chris.

Houses in the compound were older wooden prefabs, built on stilts for protection from animals and insects. Chris remembers playing in the sheltered area under the house. At times they could hear the water buffalo scraping their horns on the side of the house at night and the giraffes wandering through and destroying the clothes lines. Animals came close to the house for the fresh grass that flourished in the discharged grey water. In the distance they could hear lions and hyenas. It was a rich environment for wildlife. Apart from the generators, they would not venture anywhere on foot in the darkness. They always drove, even for short distances.

Water was a problem at the compound. The SRI was at the end of a 52-mile pipeline from a spring. The pipes were buried at a shallow depth and were frequently dug up by elephants who detected a leak. The water supply could be disrupted for days and sometimes a week or two. As a result, the SRI had to make do with very little water at times. Generally, there was never enough, and what there was had to be conserved.

Second child

Daughter Sarah was born at the SRI in December 1974. John had proposed an extension of his contract so that Joan would not have to travel during her pregnancy. There was a choice to either go to Nairobi and be stuck in a hotel room waiting for a long period, or to give birth at the SRI. The latter was preferred, as they had Chris to look after and no family support available. The spouse of a Dutch scientist was a medical doctor and was comfortable with home births. As this was the second birth for Joan, it was considered safe. When Joan's water broke in the middle of the night, John had to leave her alone with four-year-old Chris as he drove to get the doctor. He then assisted by boiling water. The birth went well and was followed by a couple of unique customs: it was a local practice to throw the afterbirth to the hyenas, and a Dutch custom to facilitate bonding by having the sibling hold the newborn within minutes.

Before Sarah's birth, Joan had usually been with John in the field, helping in the position of field assistant. She kept the camp running, took notes and made wildlife observations. "She was very much part of the research," said John. Her executive assistant training and clerical/administrative experience with scientific responsibilities were invaluable. "The FAO was getting a real bargain," said John as he praised Joan's contribution to his work.

Once Sarah arrived, however, it was almost a full-time job for Joan just keeping the family fed, especially Sarah. "I was nursing her and had a good food grinder," reflected Joan. It wasn't easy to get the foods that she preferred. Food was obtained on a weekly trip to Arusha, but fresh vegetables and tinned goods were not always available. Resourcefulness was necessary. "Comfort was hard to achieve," said John. As always, when Joan had time she assisted John with typing and report writing.

Non-work life

John was issued a Land Rover for work which could be used for personal transportation, since work blended so strongly with their personal activities. The UN advisers were helpful in authorizing trips to Nairobi that could be combined with personal grocery shopping to augment their meagre supplies from Arusha.

Despite its primitiveness, there was a country club atmosphere at the SRI, rich in cultural experiences. When they got to the coast, they sailed, swam, and surfed. Non-work life was pleasant and relaxing. From the compound there was the ever-present view of savanna grasslands with acacia trees and assorted animals and birds. The avocation of birding was extremely interesting in this part of the world. There was quite a mix of wildlife research and wildlife viewing as a form of recreation—"we were exceedingly fortunate, unusually so," said John.

SERENGETI SAVANNA GRASSLANDS IN FRONT OF THEIR HOUSE

Game cropping

The project explored wildlife utilization in a broad sense, including tourism in the form of game viewing, photography, and safari hunting; subsistence hunting; and game cropping. Particular emphasis was placed on the latter. John was involved with a detailed review of four game cropping projects and ecological investigations of each area. Again, game cropping had the taint of poaching because it was done so close to park areas. Locals were shooting the wildebeest, which came through in large numbers. These were the animals considered to be most expendable. More desirable were the lions and zebras. Their meat could be sold, and they had higher value—meat, trophy and tourism value. The project focused on lower-value animals like wildebeest, trying to determine the numbers that could be taken by local people while not harming the animal population. Locals argued that the population was so great that tens of thousands could be harvested without harm.[1]

Work was a combination of field work, reviewing files, writing up field study results and putting together proposals for wildlife utilization. There was non-consumptive use, as in tourism, and consumptive, the possible harvest of wildlife on the outside edges of the park. The work was controversial with John's colleagues, who were very conservation-oriented and didn't always understand that game cropping could be a conservation method. Tact and diplomacy were necessary. "I myself had difficulties reconciling the idea of cropping on the edge of a national park with complete protection within the park," John admitted in retrospect.

"Fieldwork was always exciting," said John, with Joan in agreement. "There were always new areas to see, new animals to learn about, and people to meet."

International collegiality

Living at the SRI among international scientists greatly expanded John's thinking about what was possible and how unusual ideas could be approached. There was a "wonderful mixture of ideas" among the British, European and American scientists. "I had the benefit of the collegiality that went with living in a community of scientists and benefitted greatly from that association," said John. This collegiality became his future reference point for how to engage in scientific discussions. Here he was even able to share his curiosity about Sanderson's ape-man report and received encouragement to pursue this interest.

Since reading the account of the ape-man in British Columbia back in his undergrad years, John had a curiosity about great apes. It was during their time in Tanzania that he and Joan had the opportunity to meet Jane Goodall, the world's foremost authority on the social and family life of chimpanzees. They were able to observe chimpanzees with her in Gombe National Park, where she had been working since 1960. Gombe National Park, established in 1968 on the eastern shore of Lake Tanganyika, is home to chimpanzees, baboons, monkeys and more than 200 bird species. John had previously observed gorillas during his time in Uganda. He was contemporary with Dian Fossey, the American primatologist and conservationist, and knew of the work she was doing documenting the peaceful nature and nurturing family relationships of the mountain gorillas in adjacent Rwanda. These pioneering female researchers, Goodall and Fossey, were revolutionizing scientific methodology in primatology by living among their subjects. John wondered about the "ape-man" in British Columbia. Could there be a great ape in North America?

CHAPTER 6: SERENGETI SOJOURN

Enter John Green

It was John Green, from Harrison Hot Springs in British Columbia, who wrote the first popular books on the sasquatch: *On the Track of the Sasquatch* (1968), *Year of the Sasquatch* (1970), and *The Sasquatch File* (1973). He was the first to try to attract the interest of relevant scientists to study sasquatches. John learned of Green's work and introduced himself in correspondence from the Serengeti. He indicated that he would like to come to BC to conduct sasquatch research, and requested to meet with him. Shortly after their return to Canada, John wrote from Ontario:[2]

June 16, 1975

Dear Mr. Green:

I was most grateful for your letter of January 30th which I received while I was still in Tanzania. I am now back in Canada and expect to arrive in B.C. between mid-July and early August... I want to get started in sasquatch research as soon as possible and plan to spend at least until Christmas in initial investigations.

Your comments on the suitability of various areas were most appreciated. I still wish to conduct interviews, etc., on Vancouver Island before dropping it as a potential study area. In this regard it will be most helpful indeed to meet you and discuss various points. Thank you very much for your invitation to drop in and see you before starting. I very much look forward to this.

Yours sincerely,
John Bindernagel
Wildlife Ecologist

In this letter, John identified himself as a "wildlife ecologist," which emphasized the practice of wildlife management, a significant aspect of his overseas work as a consultant with the FAO. John had already identified Vancouver Island as a potential area for study, based on Sanderson's writing.

Green had a master's degree in journalism and ran the *Agassiz-Harrison Advance* newspaper. His research and writing elevated the profile of the sasquatch in BC and brought him in contact with explorers and academics. He first became interested in sasquatches after hearing a story of an experience of a family at Ruby Creek in the Fraser Valley. Two years later, in 1959, he participated in "The Pacific Northwest Expedition," organized by a wealthy Texan in search of sasquatches, and met Ivan Sanderson, then a writer from New York—the author of the article that grabbed John's attention in *True Magazine* in 1963.

HIS MENTOR, JOHN GREEN
(MURPHY-HANCOCK COLLECTION)

Between the 1960s and 2000, Green archived interviews and documented detail on sasquatch behaviour, anatomy, and ecology. He welcomed John, a professional scientist, to join the sasquatch research scene in BC. He had been actively inviting the participation of academics, with slow success. Out of the blue, he had received John's correspondence and was excited by the prospects of a professional scientist with an interest in sasquatch research coming to BC. "He became a friend and mentor," said John with appreciation.

CHAPTER 7

BRITISH COLUMBIA:
DOMESTIC CHOICES

In anticipation of a cross-Canada trip from Ontario to BC, John purchased an older model bread van that had been converted into a camper. Whoever had done the conversion had painted a large "W" on the side. "Is it an older model Winnebago?" inquired curious onlookers, to John's delight, as they headed west.

Vancouver Island held an attraction. In the *True Magazine* article by Sanderson that John had taken to class back in 1963, Vancouver Island was mentioned as a fertile area for sasquatches sightings.

There was a long and rather full tradition about the sasquatches in British Columbia, and especially on Vancouver Island, where so many sightings have been recorded. Vancouver Island is enormous. It is very rugged, clothed in the densest forest, and is, even today, for the most part unexplored. What is more, it was the first part of the Northwest Pacific Rain Forest to be invaded by roads, and thus first of these unexplored regions where sightings could have been made.[1]

John had not forgotten that article and he had done his homework over the years. He was aware of some of the historic sightings on Vancouver Island referring to wild men, man-beasts, and Mowgli-like creatures and their footprints:[2]

1901	near Campbell River	hairy man-beast washing roots in water
1904	near Horne Lake	hairy wild man with long matted hair
1905	Little Qualicum	stout creature over six feet tall with matted hair
1905	Cowichan Lake	wild man sprang into bushes and disappeared
1905	Comox, French Creek,	no less than 11 persons saw wild men
1906	Horne Lake	wild man with astonishing agility
1921	Englishman's River	tall, brown sasquatch at edge of forest
1928	Conuma River	indigenous man kidnapped and kept prisoner by sasquatch
1930	near Alert Bay	sasquatch walking along beach and into trees
1942	Port McNeil	six 14.5-inch footprints found
1946	near Coombes	sasquatch ran across road in front of car
1953	Oyster River	huge creature covered in dark hair

The island had more appeal than the Harrison Hot Springs area, which was evolving as the centre for sasquatch activity in BC, with John Green in the lead. Nobody was doing research on the island.

CHAPTER 7: BRITISH COLUMBIA

By September 8, 1975, "after a prolonged tour of the island," John and family were settled in Courtenay, a mid-island community on the east coast of Vancouver Island. He wrote to Green:

> ...we travelled from Port Renfrew to Port Hardy and Alert Bay looking at both real estate and sasquatch habitat. I am now well aware of the immensity of a sasquatch search in this area and will be looking for any possible assistance in the future.

They had travelled from the Pacific Coast on the west to the northernmost populated area of Vancouver Island, and even to Cormorant Island, another small island on the east coast of Vancouver Island, where pioneer and First Nations cultures co-exist. John deemed the large extent of Pacific coast wilderness on Vancouver Island and the smaller islands of the north Salish Sea and Broughton Archipelago to be potentially rich for sasquatch research.

The van, which had served them well on their trip across Canada and while exploring the northern parts of the island, was a temporary residence at the Maple Pool and Little River campgrounds until they were able to purchase a house.

While John was keen to begin sasquatch research, there was a pragmatic concern: he needed to earn a living and provide for his young family. Where and how would he do this? Not being professionally established in North America presented difficulties. Potential employers viewed his overseas work as not relevant for Canada. In his own mind he was well-qualified based on his foreign experiences, but he had to convince employers that his skills were transferable.

Comox Valley

The Comox Valley, which includes the towns of Courtenay, Comox and Cumberland, was small, affordable, and family-friendly, with attractive seasonal temperatures. The west coast marine region has a moderate, Mediterranean-like climate in the summer and minimal snow in the winter. It is known for year-round leisure and sports opportunities and is promoted as "the recreational capital of Canada." From Courtenay it was only minutes to the mountains on the west or the Strait of Georgia on the east. John and Joan bought an older wood-frame house within walking distance to the Puntledge River, where they could swim and fish.

Strathcona Park, the oldest provincial park in British Columbia, is the backyard to the Comox Valley. It is a rugged, undeveloped mountain wilderness. Except at the highest levels, winters here are milder than in the interior of the province. Based on historical accounts and his ecological awareness, John believed the Courtenay to Campbell River corridor to be prime sasquatch habitat.

Career-wise, however, the Comox Valley was not a logical place to which to move. John should have been in a larger centre if he wanted better work options, but he had already enjoyed a successful international career as a consulting wildlife biologist, and he was prepared for a change in lifestyle. He was willing to make the trade-off to be closer to nature and able to study the sasquatch—but he was unemployed.

Up north: Fort St. John

In the spring of 1976, he accepted a job with the Fish and Wildlife Branch of the provincial government, working out of Fort St. John in the northeast corner of the province, 1400 kilometres from Courtenay. A former trading post, established in 1794, it is one of the largest cities on the Alaska Highway.

In his second year, he was under contract to the BC Environment and Land Use Committee as biologist in charge of a wildlife team involved in the environmental assessment of coal exploration and development in the northern Rocky Mountains.

Joan wasn't keen on the weather. They had come from moderate wet and dry season temperatures averaging about 25 degrees Celsius (77 degrees Fahrenheit) at the SRI to winter season sub-zero daily high and low temperatures. She had to bundle the children up and be sure to cover their faces if they went outside. In fact, at least five months of the year they experienced an average of double-digit sub-zero low temperatures. The summer months, on the other hand, were warm to hot and facilitated outdoor activities. Chris took his first year of public school in Fort St. John and remembers his dad taking him fossil hunting along the banks of the Peace River.

While John enjoyed outdoor activities in all seasons, the harsh northern climate was too restrictive and uncomfortable. They were eager to get back to Courtenay after two years in the north country.

His mentor

John Green had a daughter living in the Comox Valley, and he and John got together at times when he visited the area. Green invited him along to follow up on a sasquatch report from nearby Comox Lake soon after John and Joan took up residence again in Courtenay. Green appreciated his mentee's science credentials and encouraged and facilitated his involvement in sasquatch research.

Correspondence between the two was frequent and addressed questions related to their common interest of sasquatch. Letters from the 1970s show strong collaboration. John would ask questions like: "Any suggestions for areas worth looking at?" or "Can you suggest anyone in the area who I might contact for further reports on the island?" John Green would suggest: "A good man to get to know…" or "You might want to look up…" In reply, John's letters contained responses such as: "With regard to the leads that you gave me…" He would then give a detailed response about his contact or investigation, or "I look forward to reporting to you as my investigations get underway."

John valued and was inspired by Green. Emulating his friend, he quietly began to collect his own reports on sightings and to research everything he could find on the sasquatch. His local reputation grew as his interest became known—he was seen as an affable scientist and able presenter who enthusiastically investigated and studied sasquatches. He was a curiosity with a depth of background knowledge on the subject unparalleled locally.

Speaking engagements

Speaking engagements increased and often resulted in sasquatch reports coming forth. "Behind every sasquatch report there is a story," he would say, referring to the reason why the observer decided to come forward at that particular point in time. He maintained that only a small percentage of observations were being reported, and even fewer recorded. Eyewitness reports needed to be investigated. If someone had confidence in coming forward, he felt he owed them a "best of science" response. He felt it was imperative to respectfully listen, examine evidence, do an ecological assessment, and provide expert feedback and reassurance while answering any questions.

With only sporadic remunerative employment, however, he once again turned to overseas consulting opportunities. This time there was an opportunity to explore another part of the world—the ancient Middle East.

CHAPTER 8

IRAN:

PERSIAN POLITICS

In 1977, John took an assignment as a consulting wildlife biologist with the UNFAO in Iran, where he was involved with wildlife surveys and training nationals in wildlife management. While there had been a request from the Iranian government to model the Iranian national parks system on the American model, there was also a growing anti-Western sentiment. It was the time when the Shah's monarchy was coming to an end, to be replaced by a more fundamentalist Islamic regime. Demonstrations against the Shah developed into civil resistance, with strikes that paralyzed the country in 1978, while John and his family were there. Even wildlife management and conservation were affected by the country's revolutionary politics. The Islamic Revolution of 1978-79 replaced the 2500-year Persian monarchy.

Persian history

Initially, it was a challenge to understand the regimented etiquette in Iran. John spent time in the field with young Persian biologists, sharing about wildlife conservation. Habitat protection was high on the agenda—it wasn't just dealing with hunting, poaching, and industrialization. The red deer, similar to a North American elk, was a species at risk. It was a question of harvesting and to what extent to allow trophy and meat hunting. As an outsider, however, he could only go so far in suggesting priorities. It had to be their priorities, their management of people and policies. The implementation of wildlife management programs necessarily had to consider local cultural practices.

The family lived in a small apartment in Tehran, the capital city, where Joan was often cooped up with the kids, watching the street life below from their window. It was a wealthier, more progressive area of the city, however Joan had to keep her hair covered when she went to the market or if they were in the villages. Whenever possible, they travelled with John when he was doing fieldwork.

The arrangements were "not the best," but "workable," said John. When in the countryside they stayed in a large canvas tent that they had brought from Canada, or in guest houses. With John out in the field, Joan would be with the kids in the camp. Chris was eight and Sarah three at the time. Chris was being home-schooled with British Columbia correspondence courses, supplemented by the rich culture of travel, markets, bazaars, and Persian history. John enjoyed the cultural learning, commenting that his work "was as much cultural as professional at times."

Expat community

They were part of an expatriate community of international advisers, including Europeans, British and Americans. There was an implicit respect for the tricky Persian language and culture. Dress, conduct, and respect for authority were important. Sensitivity was necessary, and guidance was provided by more experienced Americans. There was a strong U.S. presence in the oil industry and with U.S. military bases.

The family attended an interdenominational church with many other expats. A British/Iranian couple next door and an American couple down the street became good friends. Rich Iranians often sent their children to Britain or America for advanced education. Western influence, with the swiftly expanding oil-based economy, was stronger here compared to their previous African postings. Iran had changed from a conservative and largely rural society to a modern industrial, urban society in a short time.

John described the experience of working with the other consultants as "enriching." It wasn't necessarily the formal contacts on the job but rather the informal contacts, the social interactions, that allowed him to connect in meaningful ways. There was an informal collegiality among the scientists involved with wildlife biology. In particular, a British ornithologist was studying the Siberian crane, which invited birding dialogue.

Counterpart model

Consultants needed to establish a framework before significant changes could take place. Change, however, was not acceptable to everyone. Iran was a wealthy country, and the wealthy had a large say in the outcomes and recommendations. This created challenges. Which recommendations would be accepted and not rejected? Which power held sway? There had to be a guardedness, not to suggest ideas that wouldn't work. It was better to recommend modifications to current practice rather than attempt to replace it—a conservative approach was necessary. "Their practices had evolved over centuries and basically worked," said John.

There were challenges to get the Iranian biologists to see forests as habitat for wildlife and not just trees for the low-tech harvesting of charcoal. Hardwood forests of beech, maple, and oak produced charcoal, which went to the towns and cities. Human comfort was a higher priority, and animals needed advocates. There was always the trophy-hunting issue. It was lucrative for the government to sell licenses, but over-harvesting of red deer could reduce the gene pool. Should the number of licenses be limited? There were pressures, because wealthy Iranians and Europeans wanted to hunt red deer, and North Americans came for the mountain sheep.[1]

There was an ongoing discussion regarding ownership of wildlife. Were they owned by the ruling class for their trophy hunting and food? If owned by them, people would poach. "For whom were we protecting wildlife habitat and the wildlife resource?" John reflected. He understood why there was a need for outside in-put, but being "dropped in" brought challenges. Could he bring to bear his experience from other international assignments?

Advisors were given Land Rovers for travel. In the mountains, however, they sometimes had to resort to horses and mules. Joan related how they particularly enjoyed staying in rural villages. Advisors operated with a counterpart system. John always had an Iranian biologist as a guide and cultural facilitator, a system that he said "worked very well." Villagers were "remarkably warm and helpful." John described them as "salt-of-the-earth type people." They trusted that the recommendations would be in their best interests. "I wanted to honour that trust," John said. "I wanted to make it a workable system with government." He felt that if he could not solve the problems, he could at least shed light on how to improve the situation.

Persian culture

Persian food was a highlight of the time in Iran. The family developed healthy food habits and a taste for local cuisine—yogurt, garlic, onions, vegetarian kabobs, fresh breads. Even the tradition of kneeling around a tablecloth on the ground and eating with fingers rather than utensils was enjoyable. They drank "gallons of tea." It was a ritual to drink "hot teas," which often had to be poured into a saucer to cool. They didn't want to offend, and always tried to be deferential.

Poppies grown in the mountains were not an issue. Farming and hunting were essential to existence. Irrigation was important, and efficient use was made of ditches and canals to water patches of agricultural land. John's family was on the historic "silk road" to the east, the caravan route with its well-preserved old buildings. This was an historically important international trade route from China to the Mediterranean.

John enjoyed the opportunity to engage in wildlife management in Iran. He found the blend of northern mammals and southern desert animals zoologically interesting. The family was intrigued by the Persian history, ancient mosques, carpet weaving, and the bazaars. Joan shared the sense of adventure in Iran. She underwent hardships but saw benefits that outweighed the negatives for them as a family. "It was a great learning experience for all of us," she said.

In correspondence with John Green, after returning to Canada, John wrote: "I spent a lot of time in north Iran in the Elborz Mountains, but never was able to get any indication of an ABS [abominable snowman] story or legend."[2] Even in Iran, he had the sasquatch on his mind and hoped to come across stories of the regional relative.

As this assignment drew to a close, John recalled the advice of his CIDA advisor while in Uganda. He had recommended staying in Canada and building a career before engaging in further international work. Now

John had to face a decision: Was he to continue to pursue international assignments because of the challenge and stimulation, and rewarding opportunities to serve and learn, or was it time to return to Canada and bring long-term stability to the family?

CHAPTER 9

COMOX VALLEY:
HOME AND ABROAD

Back in the Comox Valley, John began to re-establish a consulting career as a registered professional biologist (RPBio). He had a strong background to offer in wildlife biology, zoology, parasitology, ecology and conservation. Sarah was now of school age and stability was needed for the education of the children. He thought he was ready to settle down.

Wildlife consultant

Through contacts within his professional association, John was able to promote his availability for contract work again. As luck would have it, Don Blood was now working on the island. Beginning in 1979, John worked as a consultant in wildlife research and management for Donald A. Blood and Associates, Wildlife Resource Consultants, out of Nanaimo.

"I always felt a bit guilty in that I only had technician-level work for him, while he had a PhD,"[1] wrote Blood. "I used John when we had more work than usual." Mostly he was contracted to study the impact of environmental changes on wildlife and the ecosystems on which they depended. The work required John to be in the field, which he really enjoyed, often in remote areas of the province. "His lack of domestic experience wasn't of much concern to me," wrote Blood. "He was an observant field worker and kept good notes." Early projects included study of the potential impacts on wildlife resources for the Kemano Completion Project and the Kitsault molybdenum mine.

"When John went into the field I didn't have to worry much about his safety, as I knew he was very capable of looking after himself. He was suited for freelancing and would not have fit into a structured government position," Blood said.

Work, however, was never continuous. There was always uncertainty about when the next contract would begin, and John had difficulty waiting. On the positive side, contract work did allow for a flexible schedule, including time for short-term work overseas.

Africa calling

In 1980, John accepted a short-term assignment in Mozambique. However, there was a difference this time—he would be going without his family. The children were established in school and the assignment was for weeks, not years.

Again it was a UNFAO project on game cropping, focusing on buffalo. It was similar to the previous projects, with the same issues. Hunting by locals was a major threat to the buffalo. The Marromeu National Game Reserve had been established in 1969. It was situated on a floodplain of the Zambezi River, with a lot of open grassland, rivers, and streams in the delta area. Again there was exceptional birding, as the Marromeu has 735 registered bird species, many migratory. While there, he also had the opportunity to observe some unusual wildlife species like the sable antelope, an attractive and unusual antelope with scimitar-shaped horns.

John worked alongside another UN advisor, a Portuguese fellow living in Mozambique. Most of their work involved aerial surveys from a helicopter, although they had some "pretty exciting chases" in Land Rovers. "Mozambique was not well developed, and lagged behind Uganda and Tanzania," John said.

Trinidad and Tobago

The lure of foreign lands was still there, and it wasn't long before another opportunity. In 1982, he accepted a two-year UNFAO assignment in Trinidad and Tobago, the dual-island Caribbean nation off the coast of Venezuela. Trinidad and Tobago have a rich biodiversity and the world's oldest legally protected forest reserve. They were concerned for conservation in 1776. The family was on board for another journey to a new part of the world.

The project was to strengthen wildlife management, conservation and research as a contribution to the diversification of the economy, particularly in the tourism sector. Promoting conservation and utilization of wildlife resources was considered important for the benefit of present and future generations.[2]

John served as the international wildlife management expert and coordinator of the project. He found the assignment challenging, mainly because of its emphasis on coordinating the work of colleagues—co-workers who saw no need for such coordination, especially by an outsider. There were problems establishing priorities for small wildlife projects. As an outside officer, John needed to exercise tact and humility.

The experience in the Caribbean was geographically and culturally new to John and his family, after the experiences in Africa and the Middle East. It was the easiest of adjustments because most of the people spoke English. They rented a small guest cottage in a suburb of Port of Spain, the capital, and socialized within the Trinidadian hunting and conservation communities and with expats.

"We enjoyed swimming and playing tennis at a very gentrified country club of expats," John said, smiling. Chris walked to a nearby private school for children of expats and wealthy Trinidadians. Sarah was home-schooled for the first year because she couldn't get into the private school right away. The kids enjoyed beach recreation—a lot of swimming and the opportunity to go snorkelling.

John's work was always an important part of family life. Field trips were often family affairs, as the children were allowed to skip school. Chris recalled "mangrove swamp boating trips looking for manatees, visits to caves to see roosting oilbirds, night beach searches for leatherback turtles coming ashore to lay eggs, studies with visiting biologists, like mist-netting for bats or combing small islands off Trinidad for rare skinks."[3] Going with the wildlife biologist who was studying bats was a highlight for Sarah. "We'd help catch them in special nets and we'd get a closeup view."[4] For Chris, it was the night excursions to see leatherback turtles nesting. Both remember field trips to mangrove swamps in search of Caribbean manatees previously thought to be extinct.

They were introduced to new wildlife, like armadillos and large rodents, and once again enjoyed rich birding. There were over 480 known species of birds on the two islands. "My parents were always pointing out birds of interest: macaws, scarlet ibis, and hummingbirds. The latter were prevalent at our house due to the hibiscus bushes that were there," wrote Sarah.

As children of a wildlife biologist, they were recruited to engage in wildlife preservation in a personal way. Sarah wrote: "There had been a lot of rain, and the rivers had flooded over onto the road....I went with my net to rescue the fish and put them back into the river as the waters receded."

The children had a fish tank at home with guppies, which are native to the area. When rescuing more guppies to put in the river, they came across a type of eel that also went home to their fish tank.

Joan played her usual home management and executive assistant roles. There was always typing to do for John. This trip was memorable for her, as it was the first time that she got to use an electric typewriter.

By the end of two years, it was time to return to BC and resettle the family. Chris was entering high school and Sarah would soon follow. Travel hereafter would be different. This was the last trip together as a family.

Sarah, now a mother of two, paid tribute to her mother's many years of travel when she wrote:

> *My mom was an amazing traveller. I don't know how she did it with two kids. She was the organizer for packing food, clothes, arranging flights or campsites depending on where our travels were taking us...She set up tents, kept us organized, and alongside Dad taught us about birding, plants, insects, and appreciation for knowledge and learning as we travelled. I may not have appreciated it then, but I sure appreciate it now.*

Family traditions

Chris had been initiated to the Bindernagel tradition of hunting when they lived in Fort St. John, seeking ruffed grouse. He was 14 the year the family returned to Courtenay from the Caribbean. His dad bought him a shotgun and a 30-06 rifle, and together they began intermittent duck and deer hunting—"with limited success," says Chris. "I only remember Dad bringing home a few ducks over our years of hunting." He added: "I don't think he ever hunted at all after I left home."

JOAN, CHRIS, JOHN AND SARAH BINDERNAGEL - 1987

The family fishing tradition continued as well. Chris remembers fishing with his dad at Craigdarroch Beach, where John kept an authentic handcrafted First Nations dugout tied to a buoy. This was outfitted with a Briggs and Stratton inboard. Most weekend fishermen at the time were using aluminum car-toppers, but John favoured the traditional and the handcrafted, albeit with some modifications.

Since the days of his university summer job at Fort Frances, John longed to enjoy the water again. The Strait of Georgia, the northern arm of the Salish Sea, is a major navigation channel for mariners and marine life, and rich with islands and inlets, boats and beaches. It afforded many opportunities for outdoor diversions. And he knew the coast would be prime sasquatch habitat.

Youth worker

With his children in their teens John was approached by Pat Brandon, their minister's wife, to help with the youth program at St. George's United Church in Courtenay. "The young people liked and respected him," she said. Betty Thornton, who also helped with the young people, recalled John as a good recruit. "He had so much life experience, and a high energy level—he could run circles around the kids."[5] Thornton was the organizer of activities, but John was the leader with the teens. "He had a natural rapport with them and led them in rousing games and activities." But fun and games was only part of it. "John was a good discussion leader,"[6] said Brandon. "He was good at finding God in the world around us." There was so much to be discovered. "Discovery" was a theme, and he would always stop and have a discussion.

Chris describes how his dad enjoyed physical activity during the '80s, when he was still in his forties. While in Trinidad, John loved swimming and body-surfing and built rudimentary surfboards out of styrofoam panels for the kids. Back home in Courtenay, his activities with the youth group reflected the same exuberance. He encouraged participation in year-round outdoor activities, including camping, rock climbing, mountain biking, snowboarding and skateboarding. In part he was responding to the interests of the youth he was spending time with, "but he was really leading the way in most of these pursuits," said Chris. He also worked with them on human-powered vehicle construction and restoration of old Volvos, both personal hobby areas. "He thoroughly enjoyed all the activities, and loved to share them with others. My friends and I were happy to follow his lead," reflected Chris. Life was an adventure, and there was so much to be discovered.

JOHN WITH HIS FIRST SNOWBOARD (JOE ZINER PHOTO)

The accompanying picture shows him with his first snowboard, a Mistral that he kept riding into his seventies. "It was a carving board which he settled on as being more suited to an older guy, leaving the freestyle boards for tricks and jumps to the next generation," said Chris with a smile.

While he did not see himself as a teacher, he did have an impact on youth. He became a mentor to the kids, showing them that he genuinely cared about them. He is remembered more for his down-to-earth conversations and direct messages to them. "When he looked them in the eyes they had to be honest with him," said Brandon. "He was such a good man."

"He lived his faith," said Thornton, "and the kids saw him as honest and trustworthy." In short, John was a good role model for working with youth. In a word, he was "authentic."

Years later, at a reunion of youth at a wedding event, a third party unaware of John's history with a young man proceeded to make an introduction. She was surprised to hear the response: "John is the reason we are all out here hiking and canoeing."

"This was an overstatement," John said as he related this story. He continued, "but it was awfully nice to hear." While it may have been an overstatement, it was definitely a tribute to the memory of the outdoor activities that he had promoted with the youth of the church. Today the former youth remember his patience and engagement in their interests and give testimony to his influence in their lives. Kendra Strong acknowledged his "high regard for fellow sojourners," his "intellectual integrity," his "enduring curiosity and enthusiasm." She ended with:

> I've never met anyone whose pure delight in creation matched Mr. B's.... I particularly remember hiking along the trail descending from Cream Lake when we stepped over a slug. I was concerned with not getting the slime on my boots. Most of the group, exhausted, just wanted to reach the end of the trail. Mr. B made a discovery: "An albino slug! I don't believe this has ever been documented!"[7]

Short-term international assignments

After Trinidad and Tobago, John accepted only short-term international assignments, again with the UNFAO. He travelled on his own to:

Grenada (1986): Review of impacts of proposed forestry development

Nepal (1988): Evaluation of a wildlife project

Belize (1988, 1990): Wildlife surveys and staff training

Southern Africa (1990): Review of wildlife conservation programs in Botswana, Malawi, Zambia, and Zimbabwe.

However, while the overseas assignments continued to provide income, "there was too much time away from home and family," said John. The work, always interesting, was in conflict with his responsibility to Joan and the children.

Gifts from his times abroad form part of the family memories. "I remember my dad bringing a black doll wearing a pink turban and dress with red and white bloomers back from Grenada. I still have it sitting on my dresser and remember my dad when I look at it," wrote Sarah.

When John reminisced about these trips, there were always three themes: the work, the animals, and the birds. Because of the short nature of the assignments, and the lack of family with whom to share, there was less cultural impact beyond that of the natural world. The exception was Nepal, where he was very curious about the yeti, the abominable snowman. However, he was not sure how to take the responses to his general inquiries. Was he too intrusive with questions? Were people embarrassed by the yeti? Was it cultural not to share thoughts? He did not glean much information. He knew that there was so much less research on the yeti than on the sasquatch. He recalled Grover Krantz saying that the yeti is too easily dismissed, but "dominoes will fall in due course" for the sasquatch as there is so much more available evidence.

Earning an income

Short-term assignments with the UNFAO came to an end in 1990. John's contact in Rome was no longer there and there was increasing competition from younger-generation scientists for the overseas experiences. At home he still had sporadic consulting work with Don Blood. During the late 1980s and early 1990s, he was busy with projects such as studies on the: effects of logging on wildlife habitat (Queen Charlotte Islands, now Haida Gwaii), fish and wildlife habitat assessment (Quatsino), mid-coast fish and wildlife studies, aerial winter wildlife surveys (Columbia basin), effects on wildlife of ski village expansion (Whistler), effects of logging road construction on wildlife habitat (Lagoon Inlet), potential impact of a proposed pulp mill (Vanderhoof), and potential impacts on wildlife of the construction of a second transmission line between Kemano and Kitimat.

Contracts for environmental impact studies, however, were frequently funded through the logging industry. These created inner conflict: his moral obligation to protect wildlife resources was often in conflict with the funder's pragmatic expectations to remove wildlife habitat.

"He did worry a bit about working on projects that were funded by 'developers' and were in locations likely to be affected one way or another... But I remember telling John that if we weren't doing it, somebody else would be, and they might favour the wishes of the developer more than we would," wrote Blood.

However, the work he had begun with Donald Blood and Associates in 1979 was dwindling and the company was heading to dissolution. In truth, John's interest in conducting environmental impact studies was waning too, as he was spending an increasing amount of time in sasquatch research and writing.

Still in need of an income, he accepted a sessional position as an instructor at the North Island College Campbell River campus for the

1994-95 academic year. He taught a course in wildlife management. While he enjoyed interactions with students, they didn't share his enthusiasm for science. His had been a lifelong love of science that never lost its lustre and shone brightest in the outdoors, not in the classroom. In the end he knew, once again, that he was not destined to work in an academic institution.

Millennial change

The new millenium brought changes. John was 59 years old as the century turned, and he began to think of retirement. The children were successfully launched by this time: Chris with a degree in engineering and Sarah as a licensed practical nurse. Consulting work was sparse at best, and sasquatch activities demanded an increasing amount of time.

For five years, John lent time and expertise to Bigfoot Safari, a privately funded, nonprofit society dedicated to assisting with his ongoing research. Guests went out into the woods and mountainous regions of central Vancouver Island where there had been claims of sasquatch sightings, for day or week-long outings. When combined with his investigations, research, writing and speaking, it equated to more than a full-time job. Sasquatch activities dominated his life.

John struggled with balancing a lifestyle that would allow him time with family and to pursue two of his recreational interests: travel and birding.

For several summers he worked on Alaska cruise ships, where he presented slide and lecture programs on sea life, mammals (including sasquatch) and coastal life. He spent time informally interpreting industry, environmental impact, and animal sightings for passengers. Joan, always keen for a new travel experience, accompanied him on these working holidays. "The ship accommodations were some of the nicest we ever experienced in any of our travels," she reflected with a smile.

Chris and his wife, Wendy (Dyck), and their two young sons served with Mennonite Central Committee (MCC) in Bolivia from 2005 to 2008, and John and Joan made annual visits. As always, John was attentive to the plant, insect, mammal, reptile, bird and human life around him. While Joan took books for their grandsons, John prepared science experiments. For example, he set up a Baermann apparatus with the boys to have a look at the contents of Burrowing Owl pellets that they collected.

Some of the best family-time memories are associated with activities in the natural world. "We witnessed the spectacle of hundreds of marble-sized beetles converging on a bare light bulb at night while large toads gathered to lap them up as they fell to the ground," explained Chris. "We were quite excited to come across such an exciting tropical scene." As a family they always enjoyed trips into the countryside, hiking, exploring, and definitely birding.

After Chris and his family returned to the Comox Valley, Hawaii and Mexico became favourite tropical destinations for John and Joan. No seniors travelled simpler or more economically than they did. Everything they needed was in their backpacks. Accommodation was always with other backpackers in youth hostels. They enjoyed the intergenerational camaraderie of world travellers. While others their age sought the familiarity and comfort of recognized hotel chains, they delighted in the uncommon and unpredictable experiences of students on gap years, locals on public transportation, or the occasional leftover hippy from the '60s. Backpackers were a heterogeneous group who sought low-cost airlines, hostels or lowbudget accommodation, and local cultural experiences. Most important, they were students in the university of the road, majoring in life experience and enjoying the world of nature.

WINTER HOLIDAYS REQUIRED TROPICAL BEACHES

Travel kept John and Joan young at heart and youthful in appearance. In their seventies, they still carried snorkel equipment in their backpacks to enjoy the delights of tropical fish, corals, and turtles. And, of course, they took along binoculars. Birding was a constant, often *the* determining factor when discerning a potential destination. They spent more than half a century of marriage enjoying the marvels and beauty of the natural world.

CHAPTER 9: COMOX VALLEY

Sasquatch, sasquatch, sasquatch...

What happens when a very credible scientist chooses to study an animal which scientific colleagues and the general public do not believe even exists? Part Two explores John's relationship with the various stakeholder groups with an interest in the sasquatch.

As you read, consider your own position with respect to sasquatches. John would remind you to keep an open mind and carefully examine the evidence.

PART TWO

HIS PASSION

His inspiration:

"This creature has been told about by the Amerindians for centuries, and allegedly seen by white men for more than a century, and it is still being encountered today. Are we just going to let this thing slip through our fingers by sitting back and laughing it off? Here is something profoundly alive in our very midst that certainly needs proper and intelligent study, and some serious effort expended upon it. And it is a matter that might produce one of the greatest scientific discoveries of our time."

IVAN T. SANDERSON, 1960

CHAPTER 10

SERENDIPITY:
EVIDENCE FOUND

John never forgot the rough reception that he had received from his professor and classmates as an undergrad when he had raised the topic of the ape-man back in 1963. His curiosity about a possible ape-man heightened when he observed great apes in Uganda and Tanzania, and was piqued when he met John Green. His mind went wild with the possibilities in the Pacific Northwest. He hoped to prove the existence of sasquatch, initially through eyewitness accounts and evidence that others had collected. He always hoped for a clear, high-resolution photograph. However, the demands of everyday life often left his sasquatch interests at the back of his desk. It was always a matter of balancing his need for gainful employment as a wildlife biologist with his desire to conduct research on a species he had never seen. In the early years, he kept his sasquatch work to himself, not wanting to damage his professional credibility. In his own words, "I chose to be quiet because I wanted to remain employable."

Serendipity: sasquatch tracks

His interest in sasquatch dramatically accelerated, however, in the same year as his Nepal trip, 1988, when he and Joan found tracks while chaperoning Sarah's school group on a camping trip in Strathcona Park. It was a cool October day with a hard frost. The group was on a well-used trail near Helen McKenzie Lake, at about 3,000 feet, an hour's hike from the Mount Washington ski lodge, in central Vancouver Island.

Interpreting nature for high school students was a pleasurable outing for a wildlife biologist. Everything was routine, until there, in a low muddy spot, were four very recognizable and relatively fresh footprints. A creature with very large, human-like feet had crossed the trail and left unmistakable evidence. Five girls missed them, but the sixth brought them to John's attention. One might imagine John's excitement. However, he had an obligation to the group and couldn't get back for three days to make plaster casts of the tracks. Unfortunately the tracks were damaged by other hikers in the meantime and he could only get two worthwhile casts, 38 centimetres (15 inches) in length. The best one had a running shoe sole imprint on it, but otherwise the substrate allowed for a near-perfect print cast. The other had a thin heel as the weight was on the ball of the foot, resulting in a more fragile cast. He learned from this experience and hereafter encouraged others to do whatever is necessary to protect tracks until they are cast. Because the second cast was of inferior quality, he seldom showed or made reference to it.

That serendipitous event changed things. Quite by accident, he had found some real evidence for a great bipedal hominid—a sasquatch.

CHAPTER 10: SERENDIPITY

Importantly, it convinced Joan of the reality of sasquatch. "I believed in John and what he wanted to study, but finding the tracks convinced me that the sasquatch was real," she said. Before this discovery she had doubts, like many others. She had wondered if John's passion was misguided. Now they both had evidence. There could be no better way to convince her than finding tracks in the field, together.

 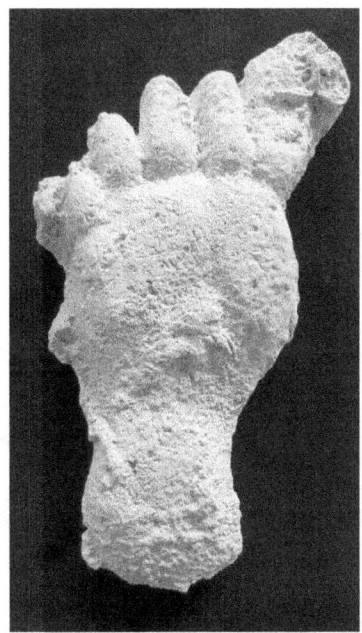

STRATHCONA PARK TRACK CASTS

Serendipity has to do with finding something of value when not intentionally searching for it. It might be described as unexpected good luck. The word was coined by a British nobleman in the mid-1700s to describe the experiences of ancient Persian fairytale characters who made discoveries through chance. If one imagines a serendipitous continuum, John's discovery would rank toward the more significant end. What were the chances of a highly interested and well-qualified wildlife biologist finding footprints when not looking for them? To John, the discovery was providential, and he had to make a decision of what to do with it.

The track cast was a gift that he was emboldened to share. This was the turning point in John's public profile. The discovery was immediately picked up by island and provincial news media and welcomed by sasquatch investigators.

Sarah acknowledged her father's excitement about the find, but admitted that as a teen, "I dismissed it quickly and went back to my social life." It wasn't until toward the end of her public school time that Sarah became aware of her dad being so involved in sasquatch research. "Even then, it didn't seem like it was all-consuming, as it did as the years went on," she wrote.

> *For me it was just part and parcel of who he was and what he did for work. It didn't seem odd to me that he researched Bigfoot and that he had a lot of info on it and talked to people about it, it was just part of life. I think when you grow up with it, it doesn't seem odd or strange or weird.*

In the acknowledgement to his first book, John thanked Joan and his children for "their forbearance" during the period of research and writing, saying:

> *In my case, greater than normal thanks are in order. Studying, speaking, and writing about the sasquatch does not confer the approval and praise awarded similar endeavours directed at large mammals such as bears and wolves. Until very recently there remained an element of embarrassment bordering on shame that one's spouse or father was not studying a real animal, but has deluded himself that he is. I am particularly grateful to my family for their confidence and support while I have undertaken this work.*[1]

CHAPTER 10: SERENDIPITY

"On my own dime"

Wildlife biologists regularly depend on tracks and other wildlife sign as evidence for the presence of bears, wolves, deer and other mammals. Tracks provide a more reliable record of a species in an area than the occasional glimpse of an animal. The tracks were a gift. His view of it was that he was simply the right person, in the right place, at the right time. But for what purpose? There had been earlier recorded sightings of sasquatches in and around Strathcona Park. John had gone with John Green to Comox Lake, on the southern boundary of the park, in 1980 to investigate calls and had followed up on sightings from nearby Wolf and Anderson Lakes.

When John had decided to pursue sasquatch research, he did so knowing there was no funding, and that he would not likely be able to attract funding in the future. Being unaffiliated with an institution or agency, there was no opportunity to apply for research grants usually available to those within academia or governmental research programs. Nobody had ever received grant money to study sasquatches because they didn't officially exist—there was no reference to them in the taxonomy of known mammalian species. At best, he hoped for a benevolent private benefactor who might support his work. This was another dream unrealized; consequently "I decided to do sasquatch research on my own dime," he said.

The expression "on my own dime" reflected back on a time when he was growing up, when a newspaper, loaf of bread, coffee, pop, or a telephone booth call only cost a dime. John could be nostalgic: things were cheaper, times were simpler when he was growing up. It was also a metaphor for his own scientific innocence, his adolescence, when curiosity was the simple currency of exchange.

Technology brought exciting opportunities for sasquatch research, but involved significant outlay of funds for computers, software, cameras, and recording devices. But there was a problem: John was old-school. His curiosity for the natural environment did not translate into an attraction to technology. Thankfully, Chris was so-oriented and was available to guide him into computer file management and video production. For assistance with book files and photo work, John had a long-valued relationship with Dennis Hiebert, a computer programmer who produced every page of *The Discovery of the Sasquatch*, along with Power-Point slides John needed for conference presentations. John readily admitted his impatience with technology, and greatly valued the assistance he received.

By the mid 1990s, John Green was into his retirement years and encouraged John, 14 years his junior, to become the "go-to guy" for sasquatch questions. John accepted the challenge, claimed BC as his bailiwick, and felt compelled to follow up on BC eyewitness reports. Proximity to Washington State quickly extended his trips throughout the Pacific Northwest.

Because he lived on Vancouver Island, there was always a cost to getting to the mainland. To keep travel expenses to a minimum, John removed the rear seats from their six-passenger family van, built a bed platform to rest over the wheel wells, and threw in a near-double-bed-sized four-inch piece of foam. There was sufficient space under the bed bench to stow things they needed to take along. Home-made window curtains were easily attached with Velcro at night. Their van became accustomed to parking in Walmart parking lots and truck stops. Meals were simple, seldom eaten in restaurants. Economy and frugality were lifestyle themes.

"It was rare for us ever to be put up in a hotel," said Joan. "Mostly, we slept in our van." Always gracious, they accepted any hospitality offered and endeared themselves to people wherever they went. "We met wonderful people as we travelled," she said.

His calling

"On my own dime" also implies independence. Assuming one has a dime, one can make choices of what to do with it. John's priorities were clear: his family, his faith, his profession. He had provided a rich multicultural experience for his family, and the children grew up with a broad tolerance and acceptance of lifestyle differences. With his children launched and his time commitments to support their activities reduced, there would now be more time for other things in life.

John always lived his life as a Christian. He felt privileged to be given the opportunity to be part of the unfolding sasquatch discovery story, and he saw this as a "calling" to which he had to respond. A calling is commonly understood as a strong urge toward a particular vocation or career. For Christians, however, it is more than this. It is a matter of discipleship. John had to ask the question: Am I doing what God would have me do? Not everyone is called to a religious vocation, but all Christians are called to offer their gifts and talents in service to others. A second question followed: Am I using my gifts and talents in a way that benefits humankind?

John believed his gifts and talents were in the understanding and practice of science. He had a lifetime of affirmation in science generally, and wildlife in particular. The call was not something from which he could walk away. When he began to think about retirement, it was really just an exercise in balancing a lifestyle so that he would be better able to respond to his call. Many go through life without experiencing a sense of call, the inward conviction that a path must be followed even at great personal sacrifice. John felt the certainty of his call.

As a professional biologist, he had a moral/ethical obligation to uphold the standards of this discipline. Even with continual rejection, he chose his words carefully. Yes, he was critical of scientists for not examining the evidence. However, he was never critical of individuals. He conducted himself with integrity. His conduct as a professional biologist was every bit as important as his conduct as a Christian—they had to be one and the same.

John saw himself as an ordinary biologist who was given an extraordinary opportunity.

Growing reputation

Over time, John's interest in sasquatches became known. He was invited to speak to groups, which often resulted in reports being shared. As his reputation grew he travelled further afield. He accepted invitations to speak at conferences, often underwriting his own expenses. He never turned down opportunities to speak to school groups, naturalist clubs, fish and game clubs, mountaineering clubs, and any other organization that invited him. He did public presentations at universities, museums, and in BC Parks.

It was an article in the *Victoria Times-Colonist* in January 1994, "Vancouver Island's little known elusive resident...the Sasquatch," which really introduced John to island readers. "Only now is he putting his name forward as a serious scientist who is looking for evidence that the large, ape-like creature called sasquatch, or bigfoot, is living on Vancouver Island."[2] The article resulted in more phone queries.

At the end of August, 1998, he wrote to Green: "I was approached by Colin Gray of CTV (W5) re sasquatch investigators. I mentioned you, Henner, Jeff, and Grover...he is also contacting Rene." *W5* was the most-watched current affairs program in Canada, which provided national exposure. The letter continued: "Lots to report...I am up to about 76 reports for Vancouver Island now. Over 250 people came out to the campground presentation last Thurs eve at Miracle Beach Prov Park (plus the *W5* cameraman and a local newspaper reporter.)"[3] John was an animated, engaging, knowledgeable presenter who easily entertained in the rustic outdoor theatre as dusk set in. Of course, there were always people who wanted to ask questions and share their experiences afterwards. He relished the opportunities to speak, always with the hope of a new story or the elusive photograph.

John eventually collected 115 eyewitness reports on Vancouver Island, many of them second-hand and lacking in complete documentation. There were times when he had to guard himself against complacency—"the ho-hum of yet another report." Some, however, grabbed his attention immediately, particularly if there was tangible evidence or detailed documentation, but many reports had little detail and were difficult to follow up.

It was the early '90s when John first contemplated writing a book. Wherever he went to speak, he was asked the same repetitive questions about the existence of sasquatches. He realized that, because of his research, he was privy to so much information that others could not possibly know. So he decided to write a book, to bring together what was known about sasquatches: "appearance, sign, food habits, and behaviour." *North America's Great Ape: The Sasquatch* was a seven-year project which he self-published in 1998 under Beachcomber Press.

During the 1990s, he became a popular conference speaker. There were invitations to present at regional Bigfoot conferences in the U.S.: Washington, Oregon, California, Utah, Idaho, Texas, Oklahoma and probably more. Joan always accompanied John to conferences, where she handled the book sales. She was amazed at how many people would come out to a conference, noting that there were much greater numbers in the U.S. "John was always excited to be at a conference," she said. With a smile and obvious pride, she continued, "People would line up to talk with him. John always listened, encouraged them, and validated their experiences." She ended with "Mostly, I was just thankful that John had some support." More than anyone, she knew his professional loneliness and the isolation that he often felt. She knew conferences recharged his batteries.

Other investigators recognized Joan's support of her husband. In a tribute to John after his passing, Derek Randles, of the Olympic Project, said "the relationship he had with his wife was second to none." In a tribute to Joan, he said: "Thank God she was able to give John his time...She was always very generous with John, and John was very generous with his time."[4]

POSTER ADVERTISING UVIC EVENT (ALEX SOLUNAC PHOTO)

John particularly liked to speak to university students, always wanting to educate future scientists and with the hope of inspiring some to follow his lead in sasquatch research. Such presentations only happened where he had a connection. For example, Alex Solunac was employed by the

University of Victoria and arranged for him to speak there. He also spoke at the University of Guelph, his alma mater, where he was remembered. And Jeff Meldrum invited him to the University of Idaho.

On the water

Part of the attraction for living in the Comox Valley was the proximity to the Strait of Georgia and the Broughton Archipelago, coastal areas that John reasoned to be rich sasquatch habitat. Since his university summer employment days on the water in northwestern Ontario, he aspired to have a boat of his own—now there was a necessity for sasquatch research.

He did eventually replace the dugout canoe with a 14-foot aluminum car-topper, which could be easily transported and launched in pursuit of the salmon run or sasquatch field work. In an email to Green he wrote: "I am planning to spend Sept. 1, 2, on the Harrison River with my 14 ft. skiff (25 hp Merc plus 10 hp back-up) looking for tracks...Perhaps you would like to join me?"[5] The following summer, after a trip to the west coast of Vancouver Island, he let Green know that he hoped "to return with my boat next week to visit Ahousat and possibly Hesquiat."[6]

When he wanted to go further afield to explore seashore sasquatch sightings on nearby islands, he graduated to a somewhat larger craft with the comfort of a cabin and the capacity for an overnighter. This boat required mooring, which he located at Heriot Bay on Quadra Island. While moorage was cheaper on Quadra, there was an additional cost associated with the ferry from Vancouver Island, and the round trip required at least three hours of travel time. Most of the overnights were spent tied to the wharf, like many liveaboards at the marina. The *Barante*, was a 1960s ChrisCraft, a quality boat in her day, with plenty of mahogany, but some of it rotten. It required constant maintenance and seldom left the wharf.

A second boat, the *Jolly Roger II* was bought to overcome the maintenance issues of the *Barante*. It was a fibreglass cabin cruiser that also remained tied up most of the time. John had dreams of trips to remote clam beaches were he could drop anchor and write while he kept an eye out for evidence of sasquatch activity. Such are the unrealized dreams of writers and wildlife researchers. The boat did become a destination for weekends away, a place to camp on the water and spend time with grandsons who explored the seashore and fished from the dock. To own a boat is part of the romance of the west coast, and John was a romantic. Not surprisingly, a family of otters became the primary residents and beneficiaries of the moorage.

Always affable, John and Joan enjoyed the camaraderie of the dock community, the live-aboards, and the fair-weather fishermen, with their nautical know-how and sea stories. Here they enjoyed an alternative, seasonal subculture reminiscent of their lost years of the '70s. They were able to fit in wherever they went. There were no pretensions—most would never know that John had a PhD and an impressive curriculum vitae.

CHAPTER 11

RELEVANT SCIENTISTS

"THEY WON'T EXAMINE THE EVIDENCE"

For John, it wasn't simply a matter of believing in the existence of the sasquatch. As a scientist, he thoroughly examined the evidence before drawing his conclusion—that there was already sufficient available evidence to document its existence. This included the track cast he made in Strathcona Park, others that he was collecting, current eyewitness reports, archival records, and photographs. While he was convinced of the authenticity of the 1967 Patterson-Gimlin film shot at Bluff Creek in Northern California, he was cautious about its use because of a wide public perception that it was a hoax. In the *Bigfoot's Reflection* documentary[1] he said:

> *I don't raise that film anymore. It's very sad, but raising it as evidence just gets you into a discussion about 'he says, she says' and claims and counter-claims. I think we should just put that film aside for now and get on with really hard evidence, especially the tracks, and then one day we can bring that film back and say that what we know about the sasquatch is there in that film.*

His position, often repeated to anyone who would listen, was: "My view is that, not only do we have sufficient evidence to treat the sasquatch as a bona fide member of North America's spectrum of large mammals, but that we already know a great deal about its biology and ecology." He particularly hoped to attract the attention of mammalogists, primatologists, wildlife biologists and cultural anthropologists.

Great-ape hypothesis

John's first book, *North America's Great Ape: The Sasquatch* (1998), did not generate the type of discussion among scientists in North America that he had hoped for. While the great-ape theory had been proposed much earlier by other writers, none had the academic credentials nor experience with the African great apes that John had. His goals in writing this book were threefold: to bring together much of what is known—or at least reported—about sasquatch appearance, sign, food habits and behaviour; to draw attention to patterns evident in these reports and to a remarkable consistency in physical features and behaviour; and to show that most aspects of sasquatch appearance and behaviour, although sometimes disturbingly human-like, resemble most closely patterns of appearance and behaviour described for the great apes of Africa and Asia—the chimpanzee, the gorilla, and the orangutan.[2]

CHAPTER 11: RELEVANT SCIENTISTS

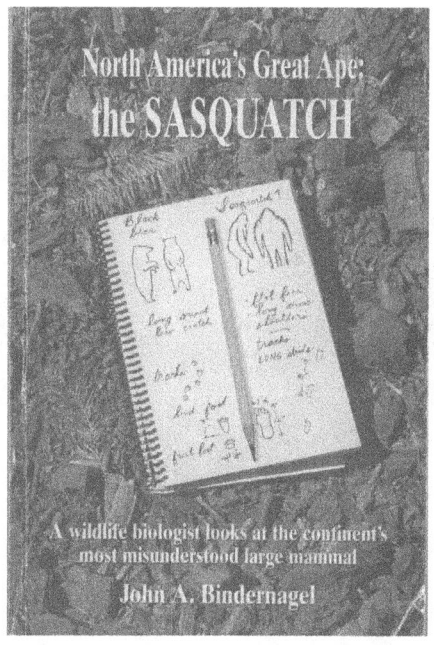

JOHN'S FIRST BOOK

Amateur investigators welcomed such an informed scientific treatment of the sasquatch as a real species, but professional scientists in North America largely ignored it. The fact that it was self-published was not in its favour. Interestingly, some high profile authorities provided very positive endorsements: Dr. Jane Goodall, British primatologist and anthropologist, the world's foremost expert on chimpanzees, author of *In the Shadow of Man, The Chimpanzees of Gombe: Patterns and Behavior*, wrote:

> *I find it exiting that, finally, a book has been written that accepts (rather than trying to prove) the existence of the Sasquatch and carefully describes the behavioral characteristics that have been recorded. The book will interest anyone who is fascinated, as I am, by one of the greatest unsolved mysteries in the natural world.*

DR. JANE GOODALL RECEIVES A TRACK CAST AND BOOK
FROM JOHN – FEBRUARY 22, 1999

Dr. Vernon Reynolds, at the Institute of Biological Anthropology, Oxford University, author of *The Apes: The Gorillas, Chimpanzee, Orangutan and Gibbon*, offered:

> *The book lays out the evidence in just the way a scientifically-minded reader would want to see it. It uses relevant data for comparisons with the Great Apes in a wholly accurate way. The result is that the readers are challenged by the many points of similarity between sasquatch anatomy and behaviour [and that of the Great Apes].*

CHAPTER 11: RELEVANT SCIENTISTS

Dr. George Schaller, author of *Year of the Gorilla* and *The Mountain Gorilla: Ecology and Behavior*, said that the book is "a fine summary of available information neatly arranged with a lot of insight and sensible deductions." And Dr. Jeff Meldrum, in the *International Journal of Primatology*, wrote: "Dr. Bindernagel summarizes over 25 years of investigation, deftly and accurately navigates the primatology literature and draws fascinating parallels that will pique the interest of the specialist and non-specialist alike."

The conclusion that the sasquatch was North America's great ape seemed so self-evident to John by the time of the book's publication that he later admitted that "expending significant further effort to attract attention to such an obvious conclusion appeared unnecessary." This he described as his own "scientific naivete."

JOHN'S SECOND BOOK

The book did generate interest among the public and was particularly well received when John was making a presentation on the sasquatch. It encouraged people to contact him with reports of tracks and sightings. Before they would talk with him, however, they wanted to know one thing: "Are you serious about this?"[3] The very fact that they would ask the question speaks to the common perception of the sasquatch as a less than serious topic.

The need for a second book became increasingly clear to John when he realized that "the discovery of the sasquatch may have occurred unannounced due to its drawn-out nature, having taken the form of a century-and-a-half long process."[4] He reasoned that a discovery claim had never clearly been made, thus the discovery had never been addressed by the scientific community. His second book, *The Discovery of the Sasquatch*, bore the subtitle *Reconciling Culture, History, and Science in the Discovery Process*.

In this book he devoted a detailed chapter to presenting the case for the great-ape hypothesis and pointed out the importance of testability for any hypotheses. He cited Kenneth Feder, who proposed setting up a series of "if...then" conditional statements (*if* the hypothesis is true, *then* the deduced facts will also be true).[5] John then gave examples of such statements for testing the great ape hypothesis:

- *If* the sasquatch is corporeal, *then* it will leave tracks in soft substrates such as mud or wet sand.
- *If* the sasquatch is a great ape, *then* it will leave tracks resembling those of a primate.
- *If* the sasquatch is a great ape, *then* it will exhibit great ape anatomical features and demonstrate elements of great ape behaviour.

To his chagrin, the great-ape hypothesis was again largely dismissed by scientists, despite its testability. On the other hand, the cultural phenomenon explanation, including the misidentified bear hypothesis, appeared to have been uncritically accepted without testing.

John wanted the great-ape hypothesis tested in the light of accumulating evidence and interpreted according to established zoological principles. However, since North American scientists believed that there were no great apes on the continent, they had no interest in testing the hypothesis. Eyewitness reports of large, hair-covered bipeds, which had left tracks across North America for more than 150 years, and increasing numbers of track casts and photographs, were not sufficiently enticing to challenge their untested assumptions and beliefs. He concluded the chapter with:

> *The possibility that great apes extended their distribution from a continent (Asia) in which at least one species of giant ape existed relatively recently (in geological time), to a continent (North America) once joined to it by dry land, does not seem so improbable—especially when it is acknowledged that other large mammals and humans have migrated between these two continents.*[6]

Dr. Leila Hadj-Chikh, in writing the Foreword to the second book, closed with: "*The Discovery of the Sasquatch* offers important insights not only about a potentially uncatalogued species, but also about the humans who have thus far declined to investigate it. Bindernagel's scholarship unlocks a door to discovery that was carelessly shut long ago, but now stands wide open, waiting for us to walk through."[7]

Book review concern

It is important for a scientist to get a new book reviewed by reputable peers in prominent journals. Journals have book review editors who consider books and solicit reviews. Once a book is submitted to a journal requesting review, the author has no input but must have faith in the integrity of the process of peer review. After encouraging endorsements from some leading international scholars for his first book, John was optimistic that his second volume would garner positive reviews.

In the case of one respected journal, however, it was not a good review. In response, John felt compelled to write to the book review editor. "I also want to thank you for taking the trouble to have my recent book reviewed, since I would rather have my work criticized than ignored," he wrote. At least with a negative review there might be something to address. In this instance there was. At the heart of his concern was the manner in which the reviewer had been chosen. John was of the opinion that the selection of a reviewer was dependent on whether or not that reviewer's own work had been addressed in the book under review, and, if addressed, how it was treated. He had been asked to review books himself and understood he should disqualify himself if his own work was mentioned in a significant way in the book. However, in this instance the book review editor had *The Discovery of the Sasquatch* reviewed by an author whose own book John had previously reviewed. John subsequently cited him in a significant way in *The Discovery of the Sasquatch* for his unsound reasoning and belittling of sasquatch research.

Being unaware of John's review of the author's book might in part excuse the editor for selecting a biased reviewer, but it did not address the impropriety of the reviewer accepting the assignment, given his acknowledged bias and John's previously unfavourable review of his book. Should he not have refrained from agreeing to review the book because he had a conflict of interest? Before registering his concerns with the book review editor, John checked his perceptions with a couple of PhD colleagues. One called the situation "ridiculous," given John's mention of the reviewer and the fact that he had previously reviewed his book; the other said it was "unconscionable," both that the editor selected a biased reviewer and that the biased reviewer did not excuse himself. The feedback he received was an encouragement to relay his concerns. As he stated: "It is not the negative review that concerns me, but that a negative review was inevitable given the selection of the reviewer."

Sasquatch: a valid claim

From John's perspective, the sasquatch as a North American mammal was a valid claim that needed to be acknowledged. This position was shared by a small number of scientists and a much larger number of amateur investigators. But John never wanted other scientists to simply accept the evidence supporting the discovery claim; he wanted them to evaluate it objectively, according to accepted scientific standards. This is the point to which he would keep returning. He wanted the dialogue, the professional interaction with colleagues to challenge the evidence and corroborate findings. There was new evidence that required the skills of other disciplines: vocalizations, hand casts, hair, twisted saplings.

But there was resistance. Most scientists wanted a cadaver and were unwilling to get involved without one. Then and only then would they consider examination of the details of anatomy, behaviours, and ecology. "My point is that we have enough to proceed without that stage. A cadaver will eventually turn up. But we do have enough information to aid us and guide us in meaningful research already," John asserted.

While a cadaver would put the issue of the existence of the sasquatch to rest, John carried a sadness, as he felt it wasn't really necessary. It was a reluctant acknowledgement when he speculated about a cadaver. There were testimonies on record of hunters who were unable to pull the trigger with a sasquatch in the cross-hairs of their rifle scope, and he pessimistically predicted that someone would come along who would have no qualms about shooting. He was not pro-kill, a position taken by some other investigators as necessary.

Alex Solunac, a long-time sasquatch researcher on Vancouver Island, described how he would not want to publicize a location if he found a sasquatch. Reading between the lines, there was a respect for the animal that has evaded mankind for so long. "A sighting, and perhaps a photograph," would be enough to satisfy his decades-long quest. Similarly for John. A cadaver would bring a flurry of scientific interest, and perhaps even research funding, but the thought of it created a painful dissonance. He knew that he would never be able to pull the trigger to bring down the first specimen. He had the utmost curiosity of a wildlife biologist, but even if a once-in-a-lifetime opportunity came to shoot and be associated with the identification of a new species, he knew he would not be able to do it. He had the greatest of admiration for the mammal that had eluded, and continues to elude, searchers.

In engaging people in discussion about sasquatches, John often asked the self-reflective question: "Am I ahead of the curve or delusional?" He then followed this with a rhetorical question about the discovery of the sasquatch: "Is this a valid position to be taken today?" To which he responded: "Yes, very much so, and has been for some time."

At the end of his 1998 book John wrote:

By now I hope that most readers have concluded, as I have, that the evidence for the existence of the sasquatch is indeed sufficient for us to move on, according the species its deserved place among the better-known wildlife species of North America.[8]

Toward the end of his life, after two more decades of relentlessly making the same point, he recognized that there probably wouldn't be closure during his lifetime. Again using the collective "we," referring to scientists, he argued:

Surely we can make it come closer by showing a willingness to examine the evidence that is available and see how compelling it is—and to force some decision making in terms of the priority of exploring this claim—making it a priority. We owe it to the scientifically-minded public and eyewitnesses to proceed, recognizing that we do lack a large piece of evidence but that we do have sufficient evidence to warrant our time, effort and expenditure of funds in this research. This has been provided by amateur investigators and we have an obligation to pick up the ball and run with it. We who are qualified in our scientific disciplines can do this.

Discovery process

In his last couple of years, John often spoke of the "prolonged process for the sasquatch discovery claim," a process that he argued has been on-going for over a century. In his conclusion to *The Discovery of the Sasquatch* he stated:

The claim of discovery is set down here simply as the inevitable result of serious scientific enquiry—a scientific interpretation of evidence which has merited greater scrutiny than it has thus far received. It is an attempt to recognize that what has been described, published, and otherwise documented for over a hundred years does indeed make sense—scientific sense—when scientists are willing to scrutinize available evidence objectively, and are willing to interpret it in the context of other scientific knowledge.[9]

John was humbled by the opportunity he was given: "A very ordinary biologist gets dropped into the middle of an extraordinary and significant discovery claim." He felt that the opportunity was there for others as well, but better qualified scientists simply declined or refrained from participation.

How is a discovery claim supposed to proceed? It is something that most scientists do not get an opportunity to enjoy; nor, because of its rarity, is this covered in the classroom. There are no *Cole's Notes* on this subject. In one of our sessions he mused:

> *I don't know how a discovery is supposed to occur. I picture it where a group of scientists come together and have dialogue on the evidence and maybe have some valid differences and interpretations but go away with having bounced ideas off each other and sort of go back to the drawing board and work on a new hypothesis or modify the hypothesis.*

He fit the matter of a discovery claim into his benchmark Serengeti experience, where he had enjoyed scientific consultation and collaboration at its respectful best.

In Part IV of *The Discovery of the Sasquatch*, John examined the discovery process at length. He pointed out how some discoveries become prolonged because of neglect or being ignored and may need to be "*re*-discovered." He made the case for the handful of scientists who, working with the great-ape hypothesis, have treated the sasquatch as a *de facto* discovery which has "occurred but has not yet been acknowledged."[10] The rejection of scientific papers, however, precluded wider dissemination of the evidence for scrutiny by scientists. One reason for the rejections in the peer review process "may have been the status of the sasquatch as an 'undiscovered' mammal." He took the position that it was time to claim the discovery of the sasquatch and framed the discovery claim with two components: (1) "that the sasquatch is an extant mammal species," and (2) "that it is a primate, more specifically, a nonhuman great ape."

Philosophy of science

In his effort to better understand the difficulties associated with acceptance of the sasquatch discovery claim, John began reading in the area of philosophy of science and was introduced to the work of Michael Polanyi, regarded as one of the twentieth century's foremost science philosophers. Polanyi's words resonated with John: "the scientific method is, and must

CHAPTER 11: RELEVANT SCIENTISTS

be, disciplined by an orthodoxy which can permit only a limited degree of dissent and that such dissent is fraught with grave risks to the dissenter."[11] Orthodoxy, the theories and practices generally accepted as correct by scientists, was an unrelenting issue for John. Orthodoxy was restrictive and did not accept his theory of the sasquatch as North America's great ape or the evidence supporting the sasquatch as an existing mammal. John knew he was a dissenter, as he held opinions at variance with those commonly held by scientific colleagues. It was reassuring to have Polanyi affirming the normalcy of what he was experiencing. Yes, he was also experiencing the grave risks.

JOHN ADDRESSED PHILOSOPHY OF SCIENCE AT THE TEXAS BIGFOOT CONFERENCE

Relevant scientists

John recalled his first meeting, as a newcomer to the Pacific Northwest scene in the early 1970s, with Grover Krantz at the planetarium in Vancouver. "You and I are the only scientists," said Krantz, an anthropologist at Washington State University who believed that the sasquatch was a descendant of gigantopithecus, having migrated via the Bering land bridge.

"Jeff Meldrum was expecting tenure so was keeping a low profile at this time," recalled John. The leading amateur investigators were his friend and mentor, John Green, and the charismatic and controversial René Dahinden.

The first scientific sasquatch conference, "Anthropology of the Unknown: Sasquatch and Similar Phenomena," was hosted at the University of British Columbia in May 1978. The conference was arranged by Marjorie Halpin, Assistant Professor of Anthropology at UBC. John Green had encouraged John's attendance in correspondence as early as 1976, when the conference was first conceived. However, there were uncertainties with John's schedule. In a letter of June 1977, his mentor wrote, "the conference is not being held until March, 1978, so you might want to keep it in mind after all." [12] However, John could not attend as he was in Iran at the time.

The conference apparently fell short of expectations and was never repeated for academics. Amateur sasquatch and Bigfoot investigators were disappointed. Richard Noll described the conference as "star studded with academia," but scientists "for the most part, though, spoke about the cultural significance and ramifications of the phenomena." This, he concluded, "was a little disappointing."[13] John Green complained, "essays written just for the sake of giving a paper, rounding out a resume and advancing a career in the ivory tower were irrelevant."[14]

When John came back to British Columbia, there was an absence of academic interest in sasquatches. Physical anthropologists had taken the early lead, but John reasoned that primatologists and mammalogists should have an interest if the sasquatch was indeed a great ape. Wildlife biologists would also need to be aware of the distribution and characteristics of this new North American species. Cultural anthropologists, with an interest in Aboriginal lore, would be important to help interpret the myths and legends and bridge the science and culture gap. He longed for other scientists, from any discipline, to step up in some way. "I feel like I am always on bended knee," he said.

In 2003, John was invited to Idaho State University by Jeff Meldrum, Professor of Anatomy and Anthropology, to speak on sasquatch, where he had the opportunity to spend three days in Meldrum's lab examining track casts and other materials. In a letter of appreciation to the Dean of Arts and Sciences, he wrote: "I can only commend Jeff Meldrum and ISU for addressing this subject [sasquatch] in a timely and appropriate manner, showing scientific leadership, which has been sadly lacking up to now in this subject area."[5] This was the type of scientific involvement for which he had long hoped. He and Meldrum became valued colleagues.

Biological versus social sciences

John described the sasquatch as "falling through the cracks." Not claimed by either zoology or physical/biological anthropology, it was consequently shunted toward cultural anthropology, bypassing any possibility for consideration by mammalogists or primatologists. Given his Great Ape hypothesis, he particularly hoped for the involvement of these scientists. The sasquatch was not considered within biological sciences and ended up, by default, within the sphere of social science, where reports would have more interest from the perspective of social history than from cultural anthropology.

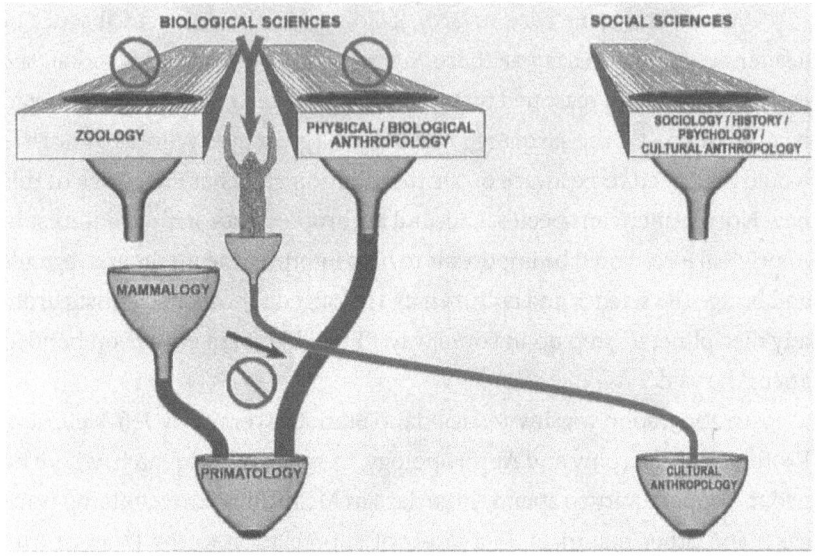

SASQUATCH FALLING THROUGH ACADEMIC CRACKS

The narrative of past events, the archival record of the sasquatch, might be of interest to an historian, but it was the cultural anthropologists who John hoped would provide opinion. However, cultural anthropologists did not show an interest. If the sasquatch was really a great ape, it had nothing to do with anthropology. "Anthropo," the prefix to their field of study, means "human."

Prevailing knowledge

From John's point of view, "prevailing knowledge" was a major roadblock. To suggest that the sasquatch was a North American great ape was far-fetched because scientists did not believe that there were great apes on this continent. Consequently, they dismissed reports of sightings of large ape-like creatures as mistaken identification of bears, which they knew for certain existed.

CHAPTER 11: RELEVANT SCIENTISTS

Or they claimed them to be hoaxes, based on a few admitted cases of men in gorilla suits or the making of plywood foot imprints. Or they accepted the stories from the Aboriginal communities as myth, because they only existed in an oral tradition and as carvings. Prevailing knowledge is what is accepted as the current truth in science.

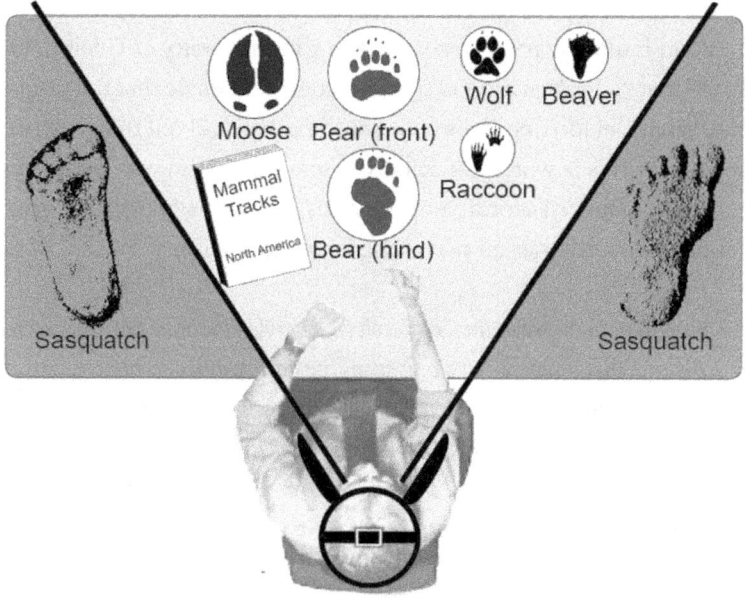

SCIENTISTS HAVE BLINDERS TO TRACK EVIDENCE

However, in the science world, knowledge is always changing. Science pushes the boundaries and seeks new understanding. It is skeptical and poses questions. It theorizes, hypothesizes, tests and re-tests. Footprints cast, trackways measured, vocalizations recorded, sasquatches photographed, affidavits sworn, and testimonies recorded are not acknowledged as evidence by many scientists. While a cadaver would change their mind, "Not having evidence you would like doesn't excuse you from examining the evidence that's available," said John, time and time again.

Tracks are an important indicator for the presence of a species to a wildlife biologist. Field guides are essential tools to assist in their identification. In the accompanying slide, John portrays the scientist as having restricted vision, limited to only those known mammalian tracks. A sasquatch track is of equal value with that of a bear or beaver; however, they were being disregarded as evidence.

In an interview prior to a lecture at the University of Guelph,[16] John asserted that "I don't go around trying to convince people that the sasquatch exists. What I'm looking for is a forum to explain and tell the evidence we have and say this is worthy of scientific scrutiny." John was trying to play the game of science according to the rules, but he wasn't getting very far. He concluded a chapter on prevailing knowledge with:

> *The strength of prevailing, or received, knowledge about the sasquatch as a cultural phenomenon—its long-standing nature and acceptance...has led to the perception that the hypothesis held by a small minority of scientists, i.e., that the sasquatch is an extant North American mammal, is speculative and unfounded. Consequently, the evidence on which this hypothesis is based continues to languish essentially unscrutinized by relevant scientists.*[17]

John usually chose his words carefully when speaking about scientific colleagues. He felt it necessary to clarify that the intent of *The Discovery of the Sasquatch: Reconciling culture, history and science in the discovery process* "was not to impugn scientists for their unwillingness to examine important evidence, but rather to understand this unwillingness."[18] Toward the end of his life, he was more blunt with his criticism:

> *It's not that scientists are out there looking for it and can't find it. Scientists are looking the other way. It's not the sasquatch which has eluded scientists,*

CHAPTER 11: RELEVANT SCIENTISTS

> but scientists which have chosen to elude the sasquatch. They don't want to hear about it.[19]

Interestingly, sasquatch study was not John's first encounter with a challenge to prevailing knowledge. While in Trinidad and Tobago with the UNFAO, conducting surveys on endangered species, senior biologists and foresters believed the sea cow had been wiped out from local waters. It wasn't until John followed a local marijuana grower into the Nariva Swamp and came back with photos that he was able to convince authorities that the West Indian Manatee still existed.

The gatekeeper effect

Annually, John took the opportunity to submit proposals to speak at conferences of the professional organizations to which he belonged. While he felt the evidence warranted presentation to his colleagues, there was resistance. The gatekeepers, individuals and committees responsible for making decisions about what was appropriate and acceptable, routinely rejected his submissions. The lack of opportunity was a frustration, and while he went through it many times, he always had the eternal hope that there might be acceptance someday.

The chairperson of one review committee justified the decision to reject a paper examining sasquatch evidence by explaining that the professional society to which the paper had been submitted

> is a very conservatively thinking group unfortunately, and as a society does not like to be associated with extreme viewpoints...Perhaps another Contributed Papers Chair would be more willing to consider your submission, but I doubt it. Until there is 'hard' evidence of their [i.e. the sasquatch's] existence the issue will remain tabloid material and not part of the scientific community.[20]

In rejecting the paper, and denying the presentation of evidence for the North American sasquatch in a scientific forum, the committee effectively contributed to the continued treatment of the subject as "tabloid material" and as a topic of pseudoscience.

Without question, the papers presented at conferences must have scientific value—they should present original ideas and new knowledge. Unfortunately, review committees may opt for the status quo, allowing for only small, incremental steps in the advancement of knowledge. A quantum leap, like the presentation of a possible new mammalian species, was simply too far-fetched to be considered seriously. Despite the mounting evidence amassed by amateur investigators and a handful of scientists, the opportunity for scientific dialogue and inquiry was routinely thwarted.

Rejection of his serious work over a prolonged period of years left John frustrated and at times angered. Was it not the responsibility of editors and review committees to foster inquiry through presentations and publications? Was his work so disreputable or far-fetched? There were times when he just needed to vent: after receiving another rejection of a paper or conference proposal, the publication of an uninformed collegial opinion, or tabloid treatment that sensationalized and commercialized the subject.

John would call and we would get together when he needed to rant. He would go on at length, in an impassioned way, with volume (due to his deafness) and animation, on scientific gatekeeping, the loss of scientific objectivity, and the ills of the hierarchy of science. Forest-bathing walks afforded John the opportunity to be heard and often led to redefining his next steps. Stopping to listen to the pileated woodpecker or observe the operations of a beaver helped to relieve stress. There was never frustration associated with wildlife observation. Like the earthen trails underfoot, they grounded him.

The gatekeepers were forestalling scientific discussion of sasquatch evidence. As a consequence, any media coverage of sasquatch reports occurred without the benefit of informed scientific comment. This resulted in the widespread acceptance of hoax claims. So much so that John concluded that "the sasquatch appears to have entered the canon of generally accepted knowledge, not as an existing North American mammal, but as a 'proven' hoax."[21]

Toward the end of his life, John frequently lamented the lack of involvement of relevant scientists: "What I find so distressing is that they seem okay with their decision to disassociate themselves from the discovery process, when they could be so helpful." In a word, he viewed this as a "tragedy." He desperately wanted the involvement of scientists, "not for my sake but for theirs, for their professional reputations and to fulfill their moral responsibility as scientists."

Problem of circularity

Gatekeepers were only one part of the problem—there was also the problem of circularity: how the unawareness of evidence and the absence of informed comment by scientists maintained the prevailing perception of the sasquatch as a cultural phenomenon.

More specifically: the sasquatch, having been categorized as pseudo-science and treated as a scientifically taboo subject, leads to:

- rejection by gatekeepers of scientific papers illustrating evidence supporting the sasquatch as an existing North American mammal, which leads to
- continued unawareness of evidence within the scientific community, which leads to
- uncritical (or default) acceptance of the hoax hypothesis along with other cultural explanations in the apparent absence of evidence suggesting otherwise, which leads to

- widespread awareness of hoaxes and acceptance of prevailing knowledge of the sasquatch as a cultural phenomenon, which leads to
- uninformed but authoritative scientific and media comment affirming the sasquatch as a cultural phenomenon, which leads back to
- the sasquatch being categorized as pseudoscience and treated as a scientifically taboo subject.

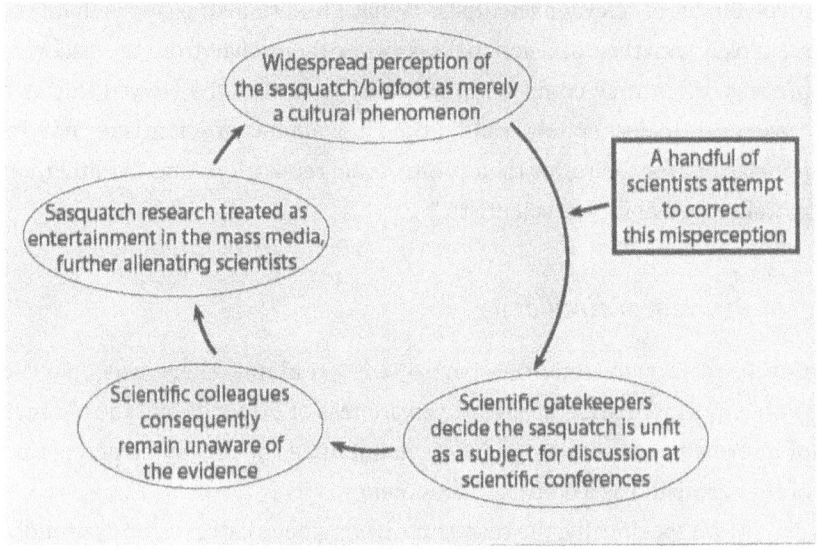

THE CIRCULARITY DILEMMA

John could accept lack of awareness of evidence as a source of scientific resistance. It led him, however, to ask two important questions: "(1) Why have scientists been unaware of the evidence? and (2) Why has this unawareness persisted so long?"[22] He acknowledged that most primatologists, mammalogists, and wildlife biologists were not aware of the evidence. As a consequence, he posited that undue reliance on authoritative opinion is often misguided, as authorities are not as authoritative as commonly perceived.

Risk to scientists

Scientists who have studied sasquatch have taken professional risks and experienced the consequences of their risk-taking. His late friend, Dr. Grover Krantz, at the University of Washington, the first academic to advocate for the study of the sasquatch/Bigfoot, did so at the expense of some promotions, and it almost prevented him from getting tenure.[23] Dr. Jeffrey Meldrum, leading American expert on the existence of giant bipedal hominids, in speaking to a wildlife science class at Utah State University, addressed the struggles associated with peer resistance. It must have been a sobering lesson as he admonished young scientists steeped in the "traditional dogmas of anthropology" not to push or to be too open about such interests. He spoke from personal experience, admitting his own naivete. "I was naive enough to jump into that deep end of the pool before I had tenure, and it nearly cost me my career," he said. He related that his promotion to full professor "was delayed by years" because of his interest in the sasquatch and he was granted tenure only "by a hair's breadth." With tenure, however, came the academic freedom to pursue his sasquatch interest.[24]

Meldrum's story illustrates the realities of a scientist in academia and one reason they might choose to distance themselves from involvement with sasquatch research. Simply put, it is self-preservation. Along with the tenure issue comes the matter of research funding. The sasquatch did not attract research money, which is so necessary for academic science careers. This is the game of science that John was unwilling to play and chose to leave back in the early 1970s.

How are young scientists to handle the orthodoxy of science when Polanyi says it "can permit only a limited degree of dissent"? In short, students of science learn not to challenge prevailing knowledge. Back in 1963, John naively challenged prevailing knowledge when he brought Sanderson's article on the ape-man to class. He could have laughed off the ape-man article the way his prof and classmates did, but it piqued his curiosity. He had been raised with scientific naivete. The classroom incident betrayed what he accepted as scientific ideals and norms. Fifty-four years later, as his life was waning, he was asked if he had any advice for young science students today. "Stay away from discovery claims that are seen as far-fetched," he quickly replied, although in jest. But then he personalized:

> *I had hoped by the end that I would find...not necessarily closure but some satisfaction in finding that it was the right thing to do, to persevere given one's convictions, which developed re the compellingness of the evidence.*

There was satisfaction in the knowledge that he had persevered in an uphill battle over the long haul. He was not completely okay, however, with the fact that there was no closure on the discovery claim. There was a sense that his life's work, his passion for the sasquatch, was left incomplete. He was still an idealist holding on to his childhood view of science, an optimist wanting to believe that things were improving. After all, the media had been coming to him for informed opinion more frequently, there were a few younger academics emerging with an interest, documentaries and YouTube were gaining important exposure, and evidence presented by amateurs was getting much better.

Authoritative opinion

John was always pleased to be able to represent science with an authoritative opinion. He quoted Polanyi, who said:

> *The overwhelming proportion of our factual beliefs continue therefore to be held at second hand through trusting others, and in the great majority*

of cases our trust is placed in the authority of comparably few people of widely acknowledged standing.[25]

Certainly the Contributed Papers Chair cited earlier in this chapter was a person of widely acknowledged standing among his scientific colleagues—and he viewed the sasquatch as tabloid material.

Ian McTaggart-Cowan, regarded as "the father of wildlife biology in British Columbia," would be one of those scientists with widely acknowledged standing both within his scientific discipline and the community at large. He established the first university wildlife program in Canada, at the University of British Columbia, the province's largest university. He said he had "serious doubts about the existence of a sasquatch...People believe in these things because they like to believe in them...And why not? It's a charming story."[26]

Grant Keddie, curator at the Royal BC Museum, took exception to John's position that scientists are ignoring the evidence.[27] He sided with scientists who said, "There just isn't any to study." He had personally conducted investigations into several sightings, including a creature running in front of a bus in the Fraser Valley, which turned out to be a hoax, a setup with a man in a gorilla suit. He maintained that, "if any evidence was to emerge that was even remotely solid, such as a bone or tooth, the top physical anthropologists in Canada would be on the next plane to B.C."

Keddie said, "when I get asked, 'Have you ever talked to anyone who is really believable or who has good solid evidence?' I say, 'So far I haven't.'"

When individuals with acknowledged standing tell their peers that they have serious doubts, that the sasquatch is just a charming story, a story suitable for tabloid publications, or that there isn't any evidence, their peers are not likely to take an opposing view. Research has shown that fitting in is more comfortable for most people than bucking convention. Even scientists are subject to conformity, as criticized by retired psychologist Thomas Bouchard:

> *The strength of this urge to conform can silence even those who have good reason to think the majority is wrong. You're an expert because all your peers recognize you as such. But if you start to get too far out of line with what your peers believe, they will look at you askance and start to withdraw the informal title of 'expert' they have implicitly bestowed on you. Then you'll bear the less comfortable label of 'maverick,' which is only a few stops short of 'scapegoat' or 'pariah.'*[28]

Bouchard points out that "conformity and group-think" attitudes are particularly dangerous in science, "an endeavour that is inherently revolutionary because progress often depends on overturning established wisdom."[29]

John coined a new term, *discovery chill*, to describe the reaction of scientists to the promotion of the discovery claim for the sasquatch. In a legal contex*t libel chill* refers to inhibition or discouragement by the threat of legal sanction. In the context of the sasquatch, discovery chill refers to the immediate rejection by scientists when discovery is perceived as delusional.

Eyewitnesses and amateur investigators appeal to authorities to explain what they have seen or found. It is precisely this appeal to him as an authority by such stakeholders that brought John the joy of explaining science and relieving their stress. At the same time, it gave him access to first-hand accounts. It was his status as a genuine, authoritative figure that also brought him invitations to participate in significant projects.

Erickson Project

Adrian Erickson, a Canadian, led a multi-year Bigfoot research project in Kentucky from 2005 to 2010. Earlier he had his own sighting of a sasquatch crossing the road in the Rocky Mountains and was determined to prove its existence. During the project, three scientists with sasquatch expertise were brought in to consult: John Bindernagel, Jeff Meldrum (anthropologist at Idaho State University), and Curt Nelson (molecular biologist at the University of Minnesota).

One of the project goals was to produce a documentary, however this idea was shelved in 2011 because, as Erickson said, he was not able to find an open-minded scientist who would stick his/her neck out to discuss the controversial DNA results, as they were not in accordance with the DNA of a likely species. There was so much opposition and skepticism that releasing the documentary at that time would have been pointless. It was surprising that, "even others in the sasquatch community in the race to come up with DNA did nothing but discredit the project."[30]

DNA analysis of blood, saliva, and hair samples indicated that female sasquatch samples come back as modern human, but those of males were an unknown species. "And here is the rub for scientists who think they already mapped out and named every species on Earth. The fact that they might have missed one is not accepted," said Erickson.

When asked by the interviewer whether he was disappointed at the reaction of the scientific community towards his project, Erickson said, "Absolutely. So much effort. We went for it all the way; did what most did not do; invested so much time and energy and dollars."

There are those who claim that the Erickson Project was a big hoax. When asked about this, Erickson replied:

The Erickson Project is and always was intended to be 100% authentic. Our team, Leila [Hadj -Chikh], Dennis [Pfohl] and myself, and our contributors Dr. Bindernagel and Dr. Meldrum are 100% upstanding, reliable citizens. The intention was to prove to the world the people coming forward with sasquatch stories are not liars, and to prove the species exists.[31]

When asked what needs to be done to get the species accepted by mainstream science, Erickson admitted, "I do not have the answer. Only scientists can explain this one."

John returned from this project believing he had seen a sasquatch for the first time. It was not a full-view sighting. The animal was partially screened by trees and moving. And while it was a meaningful corroboration for himself, it was not evidence that he could really use with others.

Russia invitation

In October 2011, John presented a paper at the International Scientific Conference on Hominology at the State Darwin Museum in Moscow, at Russian request.[32] According to Christopher Murphy, author of several books on sasquatch and curator of the Sasquatch Mystery Museum Exhibit, this was the first paper written on the ecology of the uncatalogued bipeds of Europe, Asia and North America and likely the most profound paper on John's life work.

In particular, John wanted to meet with Dmitri Bayanov, a founding member of the International Society of Cryptozoology and originator of the terms *hominology* and *hominologist*. In his book *The Making of Hominology*, Bayanov recounts how he differed with John and other North American scientists who hypothesized or stated their belief that the sasquatch was a great ape. He cited books by John Green (*The Great Apes Among Us*, 1978), John Bindernagel (*North America's Great Ape: The Sasquatch*, 1998), and

Loren Coleman (*Bigfoot: The True Story of Apes in America*, 2003), as part of the "Ape Syndrome."[33] He went on to say that "Russian hominology has never suffered from the Ape Syndrome, so prevalent and detrimental in North America."[34] He singled out John when he said, "Dr. Bindernagel and others compared sasquatches with gorillas and concluded that sasquatches are bipedal apes. Finding this conclusion incorrect, I compare homins with bipedal primates that are called human beings or people."[35] Now John had a chance to hear first-hand of Russian progress in hominology, the study of humanity's yet-undiscovered near relatives. On his return, John shared that he was thankful that they were able to talk of hominoids, a category inclusive of both apes and humans, avoiding the divisiveness of positions of the past. In his book, Bayanov acknowledged John as one of his fellow researchers from North America who "have been highly significant in my research."

Following the conference, John travelled to the region of Kemerovo in Siberia, some 3200 kilometers (2,000 miles) from Moscow, along with more than a dozen scientists from Estonia, Sweden and the United States, for a state-sponsored conference on the yeti. There had been a reported increase in the sightings of yeti in recent years.

John acknowledged that Russian authorities were promoting the organization of a yeti institute, which he and others viewed as a positive move. Later, on *The Bigfoot Tonight Show*[36] from Maryland, he said that there were two pieces of evidence that he found "very impressive" in Russia: willow and aspen saplings about 12 feet apart and arched with twisted knots. Over the years there had been lots of reports of tree modifications as possible sasquatch sign, but he had never seen anything quite like these.

John was pleased to be invited to the Russian events, which also included a day in a foot cast lab in Moscow. It was a good opportunity to dialogue with international peers.

Tipping point

John borrowed the analogy of the "tipping point" from Malcolm Gladwell, author of *Tipping Point: How Little Things Can Make a Big Difference*.[37] The tipping point is the point of critical mass or the threshold when the perception of ideas change. He combined this with E.O. Wilson's credibility scale[38] to explain the stages or steps toward the acceptance of the sasquatch as a mammal species.

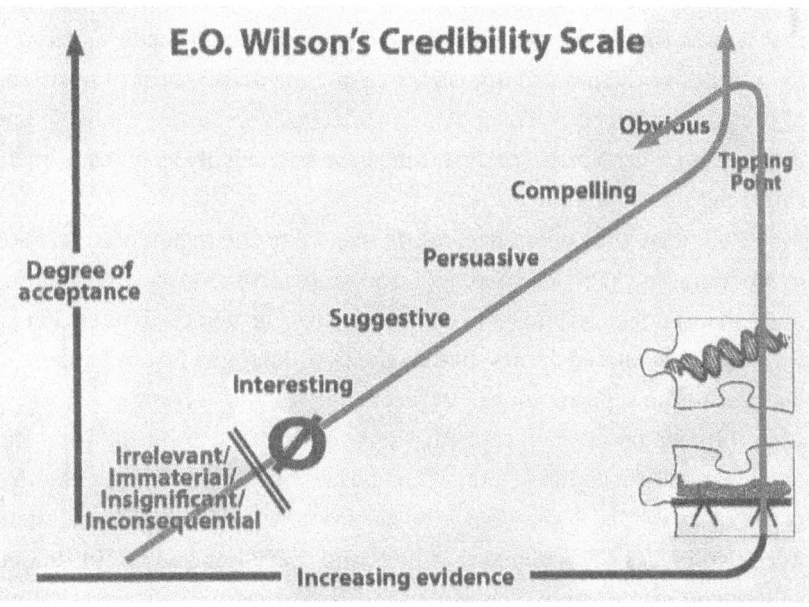

THE TIPPING POINT

On the graph he represents the archived newspaper accounts as "interesting," the sightings of a sasquatch/Bigfoot as "suggestive," the photos and videos as "persuasive," and the track casts and trackways as "compelling." Added together, they become increasingly "obvious." However, all of these stages mean nothing if the proposition is considered "unreasonable, irrelevant, or baseless" to begin with, that is to say when scientists will not look at the evidence. The tipping point, and the degree of acceptance, will be reached quickly if there is DNA confirmation or a cadaver. It is at this point that John predicts scientists will be humbled and will need to address the obvious and compelling evidence that has been accumulated.

Dissenters

John identified himself as a "dissenter," a term he gleaned from the writings of legal scholar Cass Sunstein, who argued that social pressures call for conformity and that "dissenters find themselves unpopular, or even unemployed." Notwithstanding the impact on the dissenter, "the real victims are those who are deprived of information and views that they need,"[39] he said. This certainly described John's position. As a dissenter, he felt the peer pressure to conform, as his work was not popularly accepted. However, the real victims were, as he often reiterated, his scientific colleagues and the media who did not have the benefit of the information they needed to provide informed opinion and feedback to the eyewitnesses and general public.

Sunstein describes two types of dissenters: disclosers and contrarians. John saw the small cohort of scientists pursuing sasquatch research as disclosers. They were revealing information that they had, as opposed to the contrarian, the malcontent with little constructive commentary to add. It was Sunstein's position that "disclosers should generally be prized."

The writing of *North America's Great Ape: The Sasquatch* and *The Discovery of the Sasquatch* were both motivated by the desire to disclose information, to give readers the benefit of the accumulated knowledge on sasquatch.

Interdisciplinary dissonance

In Chapter 20 of *The Discovery of the Sasquatch,* John introduces Ernest Hook's "interdisciplinary dissonance," where observations and theoretical considerations in one discipline inhibit and obstruct discovery in another. While a claim, hypothesis, or proposal might emerge in one field, it might be considered premature in another. Consequently, there will likely be rejection if a scientist from a different field is considered unaware of theoretical implications, or is deemed incapable of understanding them.

In John's view, Hook's interdisciplinary dissonance might be an important reason for scientific resistance to the sasquatch discovery claim. He explains how evolutionary biologists, paleoanthropologists, and primatologists might be expected to object to or resist the sasquatch discovery claim and concludes:

> *If Hook's concept of interdisciplinary dissonance is correctly interpreted, then there is clearly a need to more carefully address the paleontological evidence for the recent existence of a giant ape in Asia, as well as the biogeographical evidence for large mammal migrations from Asia to North America. Available paleontological and biogeographical knowledge provide a helpful perspective for more open-mindedly assessing the evidence for the sasquatch.*[40]

He went on to suggest that such knowledge might not only be "helpful," but may be "an essential component" of the discovery claim process. He acknowledged there are substantial objections to the existence of a bipedal great ape in North America, which may have inhibited scientists from scrutinizing the empirical evidence, and asserts that he, and the handful of colleagues who accept the sasquatch as extant, possess sufficient theoretical background to appreciate the enormity of the claim that they are making.

CHAPTER 11: RELEVANT SCIENTISTS

Banquet speaker

Occasionally John did receive acknowledgement from scientists for the message that he championed. He was the banquet speaker at the Washington Chapter of the Wildlife Society in 2013. The focus of his presentation was described in a subsequent newsletter:

> *Our banquet speaker was Dr. John Bindernagel, a retired wildlife biologist from British Columbia and expert on the subject of Sasquatch or Bigfoot (a controversial ape-like mammal thought to inhabit parts of western North America). Many believe this creature is mythical, others believe it really exists, and others do not completely believe either way.*[41]

John had a bell curve slide to illustrate a rough guesstimate of the acceptance of the sasquatch as an extant North American mammal. The majority are either skeptics (having doubts and not easily convinced) or agnostics (having doubts but not having studied the topic).

The Wildlife Society newsletter went on to say that:

> *Dr. Bindernagel provided an educational and entertaining summary of trends and history of reports, other information, and his thoughts on the subject. Perhaps even more relevant to us was the information he presented about how we, as scientists, think. It is important for us to be skeptical, but also to have an open mind. How each of us balances these somewhat competing interests is up to us. He discussed the processes and stages scientists go through as more acceptance of an idea occurs.*

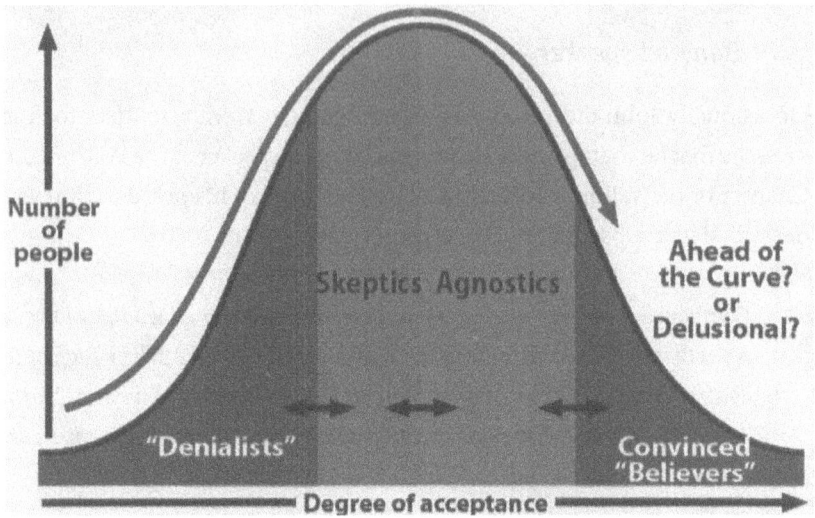

PERCEPTION OF SASQUATCH AS AN EXTANT NORTH AMERICAN MAMMAL

John was pleased to have had the opportunity to speak to professional colleagues, particularly in a forum where they were somewhat captive. They were not choosing to go to a session presented by him; rather they had paid their money to attend the banquet and would respectfully listen to the banquet speaker. While his sasquatch content was "educational and entertaining," the part of the message that was "perhaps even more relevant" was the part that dealt with how scientists handle their personal skepticism in light of new ideas. Back at home, John expressed delight at how this part of the address had been received. His philosophy of science research became an important part of his presentations thereafter.

CHAPTER 11: RELEVANT SCIENTISTS

Strategies to attract relevant scientists

As a professional biologist, John first attempted to bring his research evidence to the attention of his peers through publishing papers and speaking at conferences. These are both peer-reviewed processes, which met with little success due to the assumptions and positions of gatekeepers. In response he began to align himself with amateur investigators, in order to keep abreast of the most recent field work. His reputation grew, and he was regularly invited to speak at their conferences. His books, while garnering sales at conferences where he spoke, did not result in the sales he had hoped for. He needed another strategy for better dissemination of information.

His next step was to use technology. First he set up his own website, *sasquatchbiologist.org*. This site summarizes evidence, cultural understandings, scientific response and importantly presents his research videos. The videos are the second aspect of technology, intended to augment his two books. In his words:

> *Video presentations provide opportunities to more fully illustrate and explain documented evidence supporting the sasquatch as an existing North American mammal in the form of colour photographs, video clips, and even audio files. They also provide an opportunity to report on work-in-progress, uncompleted research projects of possible interest to other investigators. The split screen format used here facilitates narration similar to a conference presentation.*[42]

Since most of his proposals to speak on the subject of sasquatches at professional and scientific conferences were rejected, the videos constitute an available source of his research results of interest to scientific colleagues, scientifically-minded members of the public, and to sasquatch eyewitnesses seeking informed scientific comment.

The videos are also available on YouTube. The number of times the individual videos have been viewed suggests that this video strategy is working. Indeed, some testimonials received at the time of John's passing made specific reference to the videos. (For a complete list of the videos see Appendix B.) John also appears in many other YouTube videos—clips from interviews, documentaries, or conference presentations.

John put together an illustrated presentation in late 2016 for the 2017 annual conference for a society of BC biology professionals. Once again, the conference gatekeepers rejected his paper, "citing the sasquatch as a subject of cryptozoology" (considered to be pseudoscience by most scientists). He had been optimistic in submitting, as the conference theme included in part "bridging social...and scientific worlds in professional biology." From his perspective, "pointing out the necessity for increased involvement of biology professionals in sasquatch research appeared timely and relevant to the conference theme." His presentation included over 80 illustrations, some prepared specifically for the paper. In response to the rejection, John gleefully recorded the presentation as a video, titled: "A professional biologist updates his professional colleagues regarding the legitimacy and necessity of sasquatch research."

> *As such, it provides an opportunity for my professional biology colleagues to hear and see what they missed: the perspective and research results of a professional colleague attempting to enlighten them regarding an admittedly controversial scientific discovery, and one which has admittedly been appropriated as a subject of entertainment.*[43]

The video introduces a segment that was prepared and forwarded to the 2017 Beachfoot Conference in Oregon, which he was unable to attend because of cancer treatment. In this production he asserts that the absence of biology professionals from the unfolding discovery of the sasquatch

CHAPTER 11: RELEVANT SCIENTISTS

"puzzles" sasquatch eyewitnesses and amateur investigators. "In fact," he says, "it does more than puzzle them, it confuses them and it provokes questions," such as one put forth a couple of years earlier by one participant: "Why would biology professionals and other scientists squander their hard earned reputation for openness and curiosity and willingness to scrutinize evidence which challenges prevailing knowledge?"

The video goes on to update attendees on his latest efforts. While he had presented drawings of sasquatches in his two books, he asserts that the details for sasquatch need to appear in authoritative field guides. Using a template of a bear page from the Peterson field guide series, he presented a two-page spread for sasquatch identification, including: distinguishing physical features, track shapes, trackways, and a quick comparison with a black bear. The field guide entry was a purpose-driven tactic designed to appeal to wildlife biologists, members of the same professional associations to which he belonged. He acknowledged that colleagues might say that they are not ready for this yet, but countered with: "We may not be ready, but we should be ready."

SPLIT-SCREEN VIDEO PRESENTATION

As a final effort, recognizing there was little time left, John began to assemble information packages destined for museums of natural history and universities. These included: sample track casts packed in custom-cut styrofoam, a field guide, photographic evidence, and literature. The difficulty was in knowing to whom to send the information packages so that they would not be irreverently discarded. This labour-intensive effort, financed from personal funds, was underway at the time of his passing. True to his character, he was using this project as a teachable moment—teaching his grandsons how to make replicas of track casts.

Extraordinary claims require extraordinary evidence

Despite the roadblocks encountered, John was gracious toward his scientific colleagues on camera. In a CTV News program in 2017, he pointed out that there is no conspiracy, evidence is not being suppressed, rather it is just "normal scientific resistance to a discovery claim that is considered to be far-fetched."[44]

Mark Collard, a professor of archaeology and biological anthropology at Simon Fraser University, disagreed with John's position on the discovery claim and the role of scientists. In a *CBC News* article, he said that "the onus is on people like Bindernagel to apply for grants and explore the topics they feel are important and warrant further explanation." From someone inside the system, this is a typical response. "Nobody is trying to keep these people out of the mainstream. They can do what everybody else does in the field of science," said Collard.[45] But can they?

Collard went on to say that "the idea of a bipedal great ape in British Columbia is far-fetched because it doesn't line up with contemporary theories about evolution, and there is a lack of direct physical evidence of the creatures." Of course he made the usual argument: "The only really compelling evidence would be something direct, like skeletal remains or a cadaver." He maintained that scientists are constantly navigating the unknown. "That is one of the biggest drivers of scientific discovery—the desire amongst scientists to find things that are new and to disagree with the existing consensus." Isn't this exactly what John had been doing? "Extraordinary claims, it is often said, require extraordinary evidence," said Collard. One might also posit that they require an extraordinary scientist to bring them forward, to disagree with the existing consensus, to challenge contemporary theories and prevailing knowledge, and to undertake research on his own dime.

CHAPTER 12

AMATEUR INVESTIGATORS:

"WE OWE THEM"

Amateur investigators have contributed immeasurably to the discovery process of the sasquatch. They have a holistic approach, not limited in their perspectives as are scientists with their academic disciplines. They have the curiosity and stick-to-itiveness of the most ardent scientist, and while lacking in research skills they are keen to learn. Some bring skills from other disciplines that enhance the scope of inquiry. Some may have encountered a sasquatch, others have not but eagerly hope to do so.

Bigfoot Field Research Organization (BFRO)

The BFRO, founded in 1995 by amateur investigator Matt Moneymaker, advertises itself as "the only scientific research organization exploring the bigfoot/sasquatch mystery." The mission of the BFRO is multifaceted, seeking to resolve the mystery surrounding the Bigfoot phenomenon, that is, to derive conclusive documentation of the species' existence. This goal is pursued through the proactive collection of empirical data and physical evidence from the field and by means of activities designed to promote an awareness and understanding of the nature and origin of the evidence.[1]

John supported the mission of the BFRO and had the status of 'curator' with the organization. He was an original member who shared some authority and discretion to determine the credibility of regional reports and the standards of amateur researchers to qualify as BFRO investigators.

As a wildlife biologist, he was pleased with the BFRO policy "to study the species in ways that will not physically harm them." Among investigators there has always been the pro-kill advocates. His friend from Washington State University, Grover Krantz, a PhD level anthropologist, had long advocated the need for a cadaver to convince skeptics. Indeed, it is what most scientists say is necessary to attract their attention. On the other side, the anti-kill advocates argue it is not necessary; it is not scientific. John, although an experienced hunter, was on the anti-kill side. He felt there was sufficient evidence, with increasingly more being documented, that a cadaver was not an essential requirement for the study of sasquatch.

CHAPTER 12: AMATEUR INVESTIGATORS

Amateur contributions

John had praise for the contributions of amateur investigators. Foremost was John Green, affectionately referred to by some as "Mr. Sasquatch,"[2] the man who welcomed him to come to BC to study sasquatches and who became his mentor. Green brought his skills as a newspaper reporter and publisher to the field of sasquatch investigation. He was acknowledged for his leadership in raising awareness, documenting more than 4,000 sasquatch reports in a database that was unrivalled in his day, and writing the first books on the sasquatch back in the late 1960s. These were the introductory books on the subject that John read. And John cited Bob Gimlin, who accompanied Roger Patterson when he shot 53 seconds of footage of an adult female sasquatch, known informally as "Patty," at Bluff Creek, California, in 1967, for his part in this iconic amateur contribution.

Amateur investigators were the first to find and cast foot and hand prints; to describe physical characteristics, settings, and behaviours; to take photographs; and more recently to record vocalizations. "Spectograms of the latter are showing remarkable consistency—Alert Bay, BC; Norway House, Manitoba; Oklahoma," said John. "Experts are now needed to pursue the back story, examine and interpret this evidence brought forward by amateur investigators."

The interpretation of evidence is important. In some cases amateur investigators have made interpretations contrary to the views of the few scientists involved in the unfolding story of the discovery of the sasquatch. Twisted branches, for example, are touted by some amateur investigators as "conclusive" evidence, but have yet to be examined by relevant scientists for their authenticity. In order to win the interest and support of scientists, John cautioned that it is necessary to bring forth only the "best evidence" at this time. Stick structures, for example, while of interest to investigators, can too easily be dismissed by skeptics as windfall.

R.J. BANKS, BOB GIMLIN AND JOHN BINDERNAGEL
AT IBF CONFERENCE – 2017 (LORI ACORD PHOTO)

One story of an amateur investigator was on John's mind, right to the end. In bed, with memory fading in and out, he told of returning from the 2015 Sasquatch Summit Conference with an investigator who took him down a forested road in the North Cascades. Lori Simmons and her father, Donald Wallace, had been following sasquatches in the area for decades. They had an hypothesis about a sasquatch den in the root structure of a tall fir tree, from where they heard clear percussive sounds.

"We don't have to accept the hypothesis," asserted John, "but we should pursue the investigation." He was using the collective "we" of scientists, knowing that he would never be able to continue himself. He always insisted that amateur investigators needed the affirmation of relevant scientists. This was one of the last stories he told. He felt badly that he had not made more time available to return to the area and continue the investigation. "The sounds Lori recorded are real; it is the role of the scientist to identify, not abdicate," he said. "She deserves follow-up."

Citizen scientists

The concept of *citizen science* predates the 20th century. Early scientists were often gentlemen of means who could self-fund their own research projects. Today *citizen science* stands in contrast to "professional science," the work of academically qualified researchers. *Citizen science* is a term coined by Rick Bonney, an ornithologist at Cornell University, in 1995. The definitions of "citizen science" and "citizen scientists" entered the Oxford English Dictionary (OED) in June of 2014 and caught John's attention. Citizen science is defined as "The collection and analysis of data relating to the natural world by members of the general public, typically as part of a collaborative project with professional scientists." The Audubon Society's Christmas Bird Count, which John had long participated in, is a classic example of the involvement of citizen scientists to gather data that is later analyzed by professional scientists.

After the OED recognition, John viewed amateur sasquatch investigators as citizen scientists. While not having the academic training, they had the interest and desire of a scientist, similar to the enthusiasm that he had possessed in his youth. He had championed this role and contributed educational support and encouragement to amateur investigators for years without putting a name to it. He viewed Simmons and other amateur investigators as citizen scientists.

NORTHERN CASCADES – ALWAYS WITH A CAMERA
(LORI SIMMONS PHOTO)

"I was in need of assistance with field research in the Northern Cascades," said Simmons. She recalled that what she and John experienced "sounded like a gorilla throwing a massive tantrum." After the first visit John wrote to Simmons,[3] saying:

> I will be processing today's events for some time. Hope I can come up with some ideas to help you proceed...and to build on your work and your father's work. I now realize that this is long term study and I am very grateful to have been invited to participate. Hope I can contribute. Thanks again for one of the most significant days in the field I have ever had.

CHAPTER 12: AMATEUR INVESTIGATORS

John realized the need for high quality audio equipment and was able to obtain sponsorship for two wildlife acoustic recorders. "Because of John I now have over 400 hours of audio data collected," acknowledged Simmons. She considered John as her mentor, and credited him with teaching her how to record behaviours, and distinguish between a growl and chest beating sounds.

In another email John encouraged her, saying, "I remain VERY impressed with your efforts and results. I can't believe that so many investigators who you have invited to your site have not taken further action of followed up." He reiterated this sentiment later in the same email saying "I am simply aghast that there has not been any follow-up of note." The vocalizations have been heard by several invited to the site. Simmons has also documented tracks in mud and snow at the same site.

ALWAYS WITH BINOCULARS AND TIME TO LOOK AT BIRDS
(LORI SIMMONS PHOTO)

Paul Graves introduced himself to the author as a citizen scientist, a term he had acquired from John. He met John in 2009 at the Yakima Round-up. He was one of the organizers and John was a presenter. They traded track casts, and a friendship began. Graves had not gone to college and John had a PhD, but they were "colleagues" with a mutual passion for sasquatch research. Graves was doing significant documentation of stick structures and trackways in Washington state and invited John to join him in the field. "John was so nice and open to listen to me (and others) about my research that I just loved corresponding with him. He was the real deal, very kind and warm," wrote Graves.[4] He provided pictures of John in the field at two different sites: examining a possible "nest" in winter and inspecting a "stick structure" site with weaving, knot tying, and a small tepee in the summer. This latter site, high in the Cascades, was monitored by Graves with seasonal videos and photos for four years.

INVESTIGATING A POSSIBLE DEN STRUCTURE IN WASHINGTON STATE IN WINTER (PAUL GRAVES PHOTO)

CHAPTER 12: AMATEUR INVESTIGATORS

Whenever John went on a road trip, Joan went with him. They made multiple visits to see Graves. With his own recording studio in his basement, it was natural for Graves to look into, and introduce John to, spectograph analysis of suspected sasquatch vocalizations.

SUNNYSLOPE TRACKWAY FOLLOWED FOR ALMOST A MILE (PAUL GRAVES PHOTO)

Graves and John exemplified the reciprocity in a citizen-professional science relationship. "He used some of my papers and other material in his presentations and talks on numerous occasions," wrote Graves. "I am a citizen scientist and John helped me become a better one." He was saddened at John's passing.

> *I shared the Sunnyslope trackway with John and he was super excited about it and he used part of it in one of his videos he made about 'Snow Tracks.' I also got one cast from the trackway after the subject (after almost tracking it for one mile) stepped on a dirt embankment with just a little dirt sticking*

> out perfectly in line with the rest of the prints and the toes were perfect in the dirt part showing dermals. I drove to Tri-Cities to show John at the International Conference there and he was so excited to see it and thanked me for showing it to him.[5]

He concluded with "That was the last time I saw John."

Bigfoot conferences

Why would a serious scientist attend a conference of amateur investigators? For John it was the way to stay current, to find out what investigators were doing, and to affirm their work or recommend modification if necessary. Conferences provided stimulation, excitement, and the opportunity to dialogue with like-minded individuals. He was not part of a professional team or department, he had no academic work affiliation. He operated as a singleton and yearned for collegial discussions like he had experienced at the Serengeti Research Institute. Amateur conferences, and their attendees, were the next best thing.

"Bigfoot conference organizers—we owe them," said John. "They have brought together eyewitnesses, scientists, amateur investigators." His use of the term "Bigfoot" was intentional to recognize the leadership of American organizers. He appreciated their dedication. The conferences addressed subjects not addressed at scientific conferences. Importantly, they presented forums for discussion. "Discovery is unfolding without scientific contribution," he asserted, "and there is a need for dialogue." Unfortunately, there is no record of the number of conference presentations that John made. Since the early 1990s he was a popular speaker and conferences were part of his annual schedule.

Conferences were also about making new connections and creating new opportunities. In 1996 John met Todd Neiss, then an active duty soldier, at the Sasquatch Forum in Harrison Hot Springs, BC, where Neiss gave an account of his 1993 sasquatch observation near Seaside, Oregon. In 1998 Neiss founded the American Primate Conservancy, a nonprofit organization whose mission is to "produce irrefutable evidence of their [Bigfoot] existence, gain them official recognition, and afford them legal protection."[6] John spoke highly of Neiss and the work of the Conservancy:

JEFF MELDRUM, TOM STEENBURG, JOHN BINDERNAGEL, JOHN GREEN AT SASQUATCH FORUM, HARRISON HOT SPRINGS – MAY 1996 (DANIEL PEREZ PHOTO)

I am impressed with his sincerity and willingness to persevere with serious investigations on his own time and using his own funds. He has presented his results in timely and helpful reports which he has shared with his colleagues. His presentations at conferences and in the media are a credit to him and to our attempts to attract serious attention to this subject. His efforts have been important in our attempts to bring this controversial subject into mainstream biology.[6]

Neiss was respected for his serious approach, longevity in the field of investigation, and their evolving friendship.

The Conservancy sponsors an annual, invitation-only, "Beachfoot" conference for researchers, to which John was regularly invited. He passed on the first few invitations because the "Beachfoot" name did not sit well with him. However, after hearing positive reports and attending his first conference, he wrote Neiss an appreciative letter, noting that he had enjoyed the experience and had taken 35 pages of notes. The relaxed outdoor atmosphere, in tents rather than conference rooms, suited him. Thereafter he looked forward to the annual invitation to the Oregon coast and the chance to share with serious colleagues, "without the burden of having to entertain an audience."

JOHN WORKING ON ORREY INNESS' JUVENILE CAST

CHAPTER 12: AMATEUR INVESTIGATORS

It was at this same conference where Orrey Iness shared a 9.5-inch sub-adult track cast with John. "No one was telling him the value of his find," said John. "This was an important part of the zoological record." He considered this a "remarkable" juvenile track, a group which was not well-represented in foot cast collections. Without attending the conference, John might never have received this specimen, which he was given permission to re-cast for archival displays.

JOHN AND DON KEATING, OHIO 1998 (DON KEATING PHOTO)

John first spoke at the Annual Bigfoot Conference in Newcornerstown, Ohio, in 1998. The organizer, Don Keating, recalled his "bubbly personality—a scholarly man, very wise, well spoken, had his act together...open-minded." Again, there were reciprocal benefits to conference attendance.

"I was honoured that he thought enough of my work that he included some of my findings and materials in both his books," said Keating. What Keating valued most, however, was their friendship. "I tried to keep in touch with him on a regular basis, but sometimes life got in the way," he admitted.

In 2002, John was again a featured speaker at the Ohio Bigfoot Organization Conference in Cambridge. It was here that he first met Todd Prescott. He and Joan visited with Prescott in Ontario after the conference and "went out for a night of researching sasquatch," an activity that he was always keen to do. It was a noteworthy night as he was first introduced to the use of a thermal imager.

INVESTIGATORS AT THE BLUFF CREEK PATTERSON-GIMLIN FILMSITE, SEPTEMBER 2003: (L-R) JEFF MELDRUM, BOB GIMLIN, JOHN BINDERNAGEL, DANIEL PEREZ, JOHN GREEN, DMITRI BAYONOV (©DANIEL PEREZ)

CHAPTER 12: AMATEUR INVESTIGATORS

In September 2003, an International Sasquatch Symposium was held at Willow Creek, California. John and Joan, and other participants, had the opportunity to visit the location of the Patterson-Gimlin film and meet Bob Gimlin. This was a significant event for John. There was a mystique and magnetism associated with this location, the birthplace of what has become a socio-cultural phenomenon. As a wildlife biologist, he valued the opportunity to see the setting, to have his own ecological understanding of the most important sasquatch footage ever recorded.

The accompanying photo was taken adjacent to Bluff Creek, just down from the Patterson-Gimlin film site. "We were all standing around waiting for a ride back to Willow Creek, and I realized it was a golden opportunity to get a group picture with all these distinguished Bigfooters," wrote Daniel Perez.[7]

John was one of the presenters at the Texas Bigfoot Conference in 2009, the same year as the Yakima Bigfoot Round-up, which was organized to recognize Bob Gimlin, with a focus on the Patterson-Gimlin film and the best evidence from sasquatch research. This latter conference was organized by Tom Yamarone, Paul Graves, and James 'Bobo' Fay.

In April of 2011, he was a presenter at the first Sasquatch Summit in Harrison Hot Springs. The event, organized by Alex and Lesley Solunac and Tom Yamarone, invited sasquatch/Bigfoot researchers, authors and enthusiasts to celebrate the life and contributions of his friend John Green.

With the publication of *The Discovery of the Sasquatch* in 2010, John's presentations shifted to include more on the philosophy of science. After the invitation to Russia in 2011, there was interest in his international perspective as well.

JOHN SPEAKING AT THE YAKIMA ROUND-UP – MAY 2009
(ALEX SOLUNAC PHOTO)

Shortly thereafter he travelled further afield in the U.S. to attend the Proof of Evidence Conference held in St. George, Utah, and the Ohio Bigfoot Conference, both in 2012. At the Utah conference he met Tony Lombardo, a sasquatch investigator from Colorado, who was producing a short film titled OHMAWING and hoped to interview John. Tony talked to him prior to John's speaking engagement and requested an interview. "John was gracious and accepted the offer, and we agreed to do the interview at the conference,"[8] he said. Tony had met up with friends from California whom he had first met on a BFRO expedition, and with whom he had subsequently gone on research outings. Together they had the opportunity to meet and share time with John. A rock formation at the top of the hill in St George was used as an interesting backdrop for the interview.

INTERVIEW ON THE OHMAWING FILM SITE (TONY LOMBARDO PHOTO)

A Sasquatch Summit took place in Grays Harbor, Washington, in 2013. This annual conference is always the weekend before U.S. Thanksgiving Day. Johnny Manson explained how John came to the first conference simply as an attendee and how he worked out a short spot for him to give a presentation "off the cuff." After that, he was invited annually to present. In Manson's words:

> ...he brings legitimacy to this subject and honesty. That is rare. He's not trying to get 'clicks,' 'hits' or 'subscribers.' He's not forcing anything on you. It's almost like he's a really good teacher. He's teaching Sasquatch class. You trust your teacher. You can trust John Bindernagel. I don't know of a researcher that doesn't respect Dr. Bindernagel.[9]

RON PYLE, JOHN BINDERNAGEL, TONY LOMBARDO, MARK PIATTI AT THE PROOF OF EVIDENCE CONFERENCE – UTAH 2012 (TONY LOMBARDO PHOTO)

Manson recognized John as one of the pioneers in sasquatch research. And while others claimed to have the best evidence, and promoted themselves, John made no such claims and remained humble. "He gives you the information that he has, adds some possible questions and possible answers, and that's it. I would say he's one of the most professional sasquatch researchers we have had."

What was billed as the first International Bigfoot Conference (IBC) took place in 2016, in Kennewick, Washington. John was there as a presenter, along with Jeff Meldrum, Loren Coleman, Ron Morehead, Derek Randles, and others. The purpose of the IBC "was to provide an opportunity for up and coming researchers to share the stage with some of the biggest names in the world of Bigfoot research."[10] By this time, John was recognized as one of the big names. The focus of the conference is not to persuade attendees one way or the other, but rather to show the evidence that has been collected by researchers. This approach is consistent with John's personal position—all he ever wanted was for people to examine the evidence, which he believed would speak for itself.

Support groups

Many eyewitnesses contacted John to satisfy their own need to understand what they had seen and to process their traumatic experience. "Anyone who accepts the existence of the sasquatch becomes a member of a much-needed support group," he said with a smile. "It shouldn't have fallen to this level. There are no support groups for people who have seen cougars or bears." Again, he made the point that this would have been unnecessary if the discovery claim for the sasquatch had unfolded differently. To maintain the integrity of science, John always listened and provided an explanation. Eyewitnesses valued this support. Personal experiences, while burdensome to some, became the unifying characteristic of an informal support network for many.

There were other groups, however, which existed *in the hope of seeing* a sasquatch. Rather than focusing on understanding and coping strategies, the latter focus on sharing knowledge and research skills and undertaking expeditions. John considered himself fortunate to be associated with such a group in Victoria. Every time he and Joan left Vancouver Island, they arranged to go through Victoria, more than a three-hour drive to the south. It would have been more direct to go via the Departure Bay or Duke Point BC Ferries, only half the driving distance, but they liked to meet with the group in Victoria at least one way on their trip.

It was a reciprocal relationship. John was able to keep abreast of amateur research interests, use the group as a sounding board, and at the same time support their interests and provide guidance on field skills. This group is a Bigfoot Field Research Organization (BFRO) affiliate and organizes their own field trips where they use GoPro cameras, heat-sensing thermal cameras, and night vision scopes, a spotting scope with DSLR camera adapter, and car dash cameras. Of late, they have been setting cameras and bait spots on well-used trails and "call-blasting." The latter attempts to attract and record vocalizations to confirm the presence of the unknown species.

Alex and Lesley Solunac, Wayne Humphrey, Dave Hill, Steve Gray and Tyler Croft had been working together for a decade; two of them are BFRO investigators. Whenever John and Joan went to Victoria, the group got together for a pizza night. John was described as "a vortex of information." This was an apt description of his enthusiasm, excitability and animation—each visit was a whirlwind of opportunity. John would update the group with the latest sightings from Vancouver Island and what was happening elsewhere. "Meeting with John was uplifting for all of us,"[11] said Hill. He described how it is wearing to deal with so many people with entrenched beliefs and resistance to scientific evidence. "John shouldered much more of that burden, as he was such a public figure," he said.

Hill, with a degree in geography and career with the Canadian Forest Service, had met John in 2007 on a week-long BFRO expedition in the Cowichan Valley. Every night, small groups would go somewhere in the valley. "Every group, except the one I was in, would report some type of sasquatch activity...John approached me and asked if he could join me for the remainder of the expedition." Hill asked, "Why would you want to come with me? Every other group seems to have activity except the one I am in."

THE VICTORIA GROUP: DAVE HILL, STEVE GRAY, JOHN BINDERNAGEL, ALEX SOLUNAC (ALEX SOLUNAC PHOTO)

"Exactly!" exclaimed John.

The two of them shared science backgrounds and were able to discuss things objectively and established a friendship. "John and I enjoyed many exchanges, and I was very honoured to have helped him with the maps in *The Discovery of the Sasquatch*," said Hill.

Solunac, a university media analyst, interested in sasquatch since he was six years old when his mother read to him from John Green's books, was the initiator of the group. Gray, with degrees in biology and plant ecology, and similar forestry experience, joined the group in 2009. "I have yet to have a significant sasquatch experience, and yet keep optimistic that someday I might. There is about one sighting a year on Vancouver Island, so there's a chance."[12] As with members of other groups, he expressed his appreciation for the camaraderie: "I certainly enjoy getting out in the wilds of Vancouver Island and enjoy the company of my fellow sasquatch researchers," he wrote.

Amateur territoriality

Some amateur investigators are known to be competitive and self-protecting in their approach to evidence-gathering and tracking of sasquatches. They have a reputation to maintain, and exercise a certain territoriality. While John experienced a wonderful level of cooperation and relationship with most amateur investigators, there was one with whom he had difficulty back in the 1980s. In a letter to John Green, he shared a conference experience in which the well-known amateur investigator confronted him publicly:

> ...our esteemed colleague...dumped at length on me and my work. Fortunately, because of my own previous experiences with him and then yours of last year, I realized at some subconscious level that this was the ultimate endorsement and I should embrace it. So I made a few comments in response and later took him on privately, more than I thought I would. I found that most participants thought I had been unfairly attacked and also came to realize that everyone 'has his number' and sees that his attacks are essentially a response to perceived threat.[13]

It is hard to be unfairly attacked, particularly in a public venue. John, always a peacemaker, was pushed to the limit on more than one occasion by the same individual. He always made the effort to deal with the criticism one-on-one, in private. He recognized that the other individual had a problem—in the presence of scientific expertise, he felt threatened. Thus the best defence became a good offence.

John went on to say that "fortunately there are people present at these gatherings who see through the cheap shots and thoughtless criticism and appreciate those of us who stick our necks out." Despite his annoyance, John never spoke publicly about these incidents. In fact, he did not want the individual's identity revealed. On a *Sasquatch Chronicles* tribute show to honour John after his death, speakers commented that they "never heard John say anything bad about anyone."[14]

Bigfoot culture

John had some reservations about the amateur "Bigfoot" culture popularized in the U.S. His primary concern was that it is a culture of entertainment, particularly via television, which detracts from engaging in thoughtful scientific discussions. Because of the entertainment perspective, he saw serious scientists steering away from Bigfoot association, not wanting to put their careers in peril. "They simply let some evidence go," said John, "because of the risk, and some of the evidence might have warranted serious investigation."

In tandem with entertainment was the commercialization, the business aspect of sasquatch product promotion. This he described as "trivialization," the reduction of sasquatch/Bigfoot to nothing more than an economic commodity. While this is not inherently wrong, it was just not of interest to John. He was not a collector of things sasquatch, other than bona fide evidence. He championed sasquatch as an uncategorized species of great importance and worthy of scientific investigation; retailers, on the other hand, saw it as mass-produced dolls, T-shirts, and paraphernalia. Indeed, there are 32 Bigfoot products listed in a distributor's catalogue from Seattle.[15] As one of John's colleagues commented to the author: "In the old days when you were invited to speak at a conference the speaker had to write a paper, to present their evidence. Nowadays it is all about vendors and selling Bigfoot stuff: dolls, trinkets, etc...I feel they have become way too commercialized."

A third aspect was the quest of Bigfooters to "bag" a sasquatch. As a wildlife biologist, John had spent years in foreign countries, concerned with the preservation of wildlife. Now at home he had the greatest of respect for a species that had eluded man for so many years. He was concerned with ecology and the protection of habitat for the preservation of the species. With all the modern hunting advantages—scopes, night-vision goggles, drones—"it will only be a matter of time," he lamented, "until someone kills one."

Documentaries

At age 13, Darryll Walsh read Peter Byrne's book, *The Search for Bigfoot: Monster, Myth or Man*, fuelling a lifelong interest in Bigfoot. Later, as a writer and producer for television, he proposed a documentary on Bigfoot. "I had taped and watched every Bigfoot doc over the years," he wrote.

When asked why he wanted John in the documentary, Walsh explained that, despite coming across his name in articles and books, he could only find one previous documentary that John was in, and he didn't even have a speaking part. Walsh thought John was "grounded in reality and sensible, as well as being a scientist, so he would add some gravitas to the film."[16] *Bigfoot's Reflection*, a Canadian film, was released in 2007.[17] Since then John has appeared in many documentaries for television (see Appendix C: Filmography).

John accepted invitations to go into the field to examine evidence whenever he could, even when it might be controversial. He and Meldrum both appeared in the *Discovering Bigfoot* (2017) documentary. Meldrum, a university academic, admitted that he "received some criticism" for his decision to appear, but defended his decision, saying, "it is important to investigate and vet all possible sightings."[18] He and John were like-minded on this point.

Todd Standing, the producer, was sometimes controversial in his approach to sasquatch research and had both supporters and dissenters. "Frustrated and tired of having his evidence discounted," and wanting to "protect the species," he filed a civil lawsuit in BC Supreme Court in which he accused the BC Ministry of Environment and BC Fish and Wildlife Branch of "dereliction of duty pertaining to the interests of an indigenous wildlife species." Recognizing the boldness of his actions and the difficulties associated with such petitions, Standing knew that he would only get one shot at this attempt to get his evidence before the court.

"What happens if he is not successful?" posed John. Standing was willing to take risks in a manner others might not feel comfortable about. "In so doing, he might do irreparable harm to the discovery process," asserted John. Sometimes John walked a tightrope—neither wanting to offend nor deny an opportunity, nor to jump in carelessly or unprofessionally.

The judge ultimately ruled that Standing had "no reasonable cause" to sue BC for not recognizing the evidence.[19] "Sasquatch tracker's lawsuit tossed by BC Supreme Court," read the CBC headline.

John appeared in *Wildman: My search for Sasquatch* in 2016. This 26-minute interview presents an overview of his involvement in sasquatch research. He begins with how the sasquatch was a scientifically taboo subject in 1963 and illustrates how it still is today. Along the way, he explains how he always thought he would come back from his international assignments to find some young biologist taking on the challenge of the sasquatch, but that none had. He confesses how he is "running out of time" and reiterates the obligation of science to follow up with reports and give informed opinion. At the end he is asked: "Do you think we are close to finding solid proof of its existence?" In response he repeats yet again: "My point is that the sasquatch has been discovered, but the discovery has not yet been acknowledged."

While the media may have recognized John in Canada, the public hadn't. He commented on the difference in his reception south of the border: "Down in the U.S. people come out and say 'I've been following you in documentaries. Thank you for what you're doing'...but in Canada—almost never." The last sentence trailed off and was accompanied by a dismissive motion of the hand. It is probably true. Most people in the Comox Valley where he resided do not know of John's work, giving credence to the biblical adage that it is difficult to be a prophet in your own land.

Phone-in shows

Despite difficulty with his hearing, John enjoyed phone-in shows where the listening audience could ask questions. In preparation for his guest role on *Sasquatch Chronicles,* John said "The fans of *Sasquatch Chronicles* have some of the best questions I have ever been asked. Please ask your audience to feel free to ask questions and I will try to address them."[20] The questions, dialogue, sharing of knowledge was a meaningful exercise for him. In the same manner as he did when fellow students asked him a science question that he couldn't answer in high school, he pursued an answer to satisfy them. The sharing of knowledge was a cornerstone of John's philosophy. He was not bound by employer restrictions, tenure track competition, or department censure. He was completely free to be his candid self. This is how he related on the phone-in shows, endearing himself to so many people.

"Wes has broken through," affirmed John in speaking of Wes Germer and the *Sasquatch Chronicles* experience. As a result of his appearance on Germer's program, John received communication from a university professor who had tuned in to the show. The program had attracted the attention of an academic, something John had been unable to do by sending complimentary copies of his books to academics at universities.

Book publishing

While scientists write academic papers that are peer-reviewed, amateurs are more likely to contribute a book. A review of the book catalogue for Hancock House, the leading publisher of sasquatch titles, suggests that approximately 90 percent of the 30 titles are written by amateur investigators. These books reach a broad popular audience, not the audience that would be reading scientific articles in professional journals. They contribute a wealth of information about sasquatches to the general reading public.

While supportive of individual investigators, John decried the often generalized portrayal of the sasquatch based on very little evidence that had not been scientifically examined. With the popularity of self-publishing there are no standards, as demanded by editors from established publishing houses. Indeed, there may not even be an editor. John had published two books under his own imprint of Beachcomber Books. He knew the pitfalls of self-publishing.

Always the scientist, he wanted all reading to fairly and accurately portray the sasquatch. In discussing conformity and dissent,[21] he made reference to two recent sasquatch titles in which the authors demonstrated "disdain of the subject." While he acknowledged that both books "provided a service in documenting how the sasquatch has been presented in the mass media and how it has been represented (and misrepresented) at some of the early 'Bigfoot conferences' of the late 1990s and first few years of the 21st century," he felt they "may have chilling effect on disclosers already sensitive to being perceived as dissenters," and scientists who might be intrigued by the subject "may be justifiably concerned about hoaxes and hoax claims." As such they did not help the cause he was championing. Neither did he help their cause. His work was mentioned in only one sentence in one of the books.

The Olympic Project

John was pleased to be connected with The Olympic Project, "an association of dedicated researchers and investigators, biologists and trackers committed to documenting the existence of sasquatch through science and education."[22] He was listed on their web-site as a guest speaker and as a "Bigfoot legend."

The goal of the Olympic Project, "to obtain as much empirical evidence as possible in order to be prepared for when species verification occurs," mirrored John's approach. He applauded their studies conducted in non-invasive ways, with respect and sensitivity to the probable sasquatch habitat.

John's involvement with the Olympic Project boiled down to expedition time, and "being there for us when we need to ask him questions,"[23] wrote Derek Randles. "We are not scientists, so it was great to be able to reach out to him when we were stumped with anything." Despite his education, John was able to "dumb down his conversation and enable himself to talk to anyone." He was praised for his role with expedition clients, "making time to talk and visit with each and every one of them."

Perhaps the most exciting project for John was involvement with the Skookum Cast, a 400-pound partial body cast measuring 3.5 by 5 feet, which was found during a September 2000 BFRO expedition to the Skookum Meadows area of the Gifford Pinchot National Forest in Washington state. John was part of the team invited to assess this cast. Others included Dr. Jeff Meldrum (anatomy and anthropology), Dr. Grover Krantz (physical anthropology), Dr. Ron Brown (exotic animal expert), and John Green (journalist and sasquatch author). John helped with the cleaning process and hair removal. According to Meldrum, "The unanimous consensus was that this could very well be a body imprint of a Sasquatch."[24]

The Skookum Cast illustrates the best of amateur-professional collaboration. "I remember being star-struck by the academics occupying my abode,"[25] wrote Randles. "Having all my heroes under my roof was a bit overwhelming, but very cool." With reference to John, he wrote: "I remember him being very excited about the discovery. They all were. It was a fun time to be involved in research." New discovery, the heartbeat of science, was exciting for all involved. For John, it was also the collegiality and discourse.

Home visits

Operation Sea Monkey was a foray into the Canadian coastal wilderness in search of sasquatches/Bigfoot by the American Primate Conservancy in September 2016. Todd Neiss assembled a team of experienced researchers, including Ron Morehead, Thomas Steenburg, and Gunnar Monson, who headed to Campbell River to rendezvous with Tom Sewid, en route to the Broughton Archipelago Provincial Park area. This is British Columbia's largest marine park, located near the north end of Vancouver Island, with dozens of uninhabited islands and islets. On the way, the group, out of respect, stopped in to see John at his home in Courtenay. "John had been suffering with medical issues, or I am sure he would have joined the team,"[26] wrote Neiss.

John and Joan were ready for the visit and set out refreshments and a number of track casts and photos for the team to review. Ironically, John had recently been north on Cormorant Island to investigate possible sasquatch vocalizations. Years earlier, in 1975, he had identified the many islands of the Broughton Archipelago area as rich sasquatch habitat. The team stayed about an hour before continuing north to Campbell River.

RUSSELL ACORD VISITING WITH JOHN AT HOME IN COURTENAY
– SEPTEMBER 24, 2016 (KELLY ACORD PHOTO)

Russell and Kelly Acord, from the IBF, also visited at this time. As John had been unable to present at the 2016 IBF Conference, Acord arranged for a visit and a video interview. This video, dedicated to John's memory, is available on YouTube ("Dr. John Bindernagel at his home in Courtenay, BC").

Last video

John posted his last video, on track evidence, on February 9, 2017.[27] He emphasized the importance of tracks and described his latest efforts at preserving track casts. In the last year he had purchased three casts from a deer hunter near Sayward, on northern Vancouver Island, which had been cast back in the '90s. He thought that he was purchasing four, but one had been broken and discarded. He did not fault the deer hunter who had kept them in his shed for so many years; rather, he lamented the fact that "No one was telling him this was important scientific evidence."

CHAPTER 12: AMATEUR INVESTIGATORS

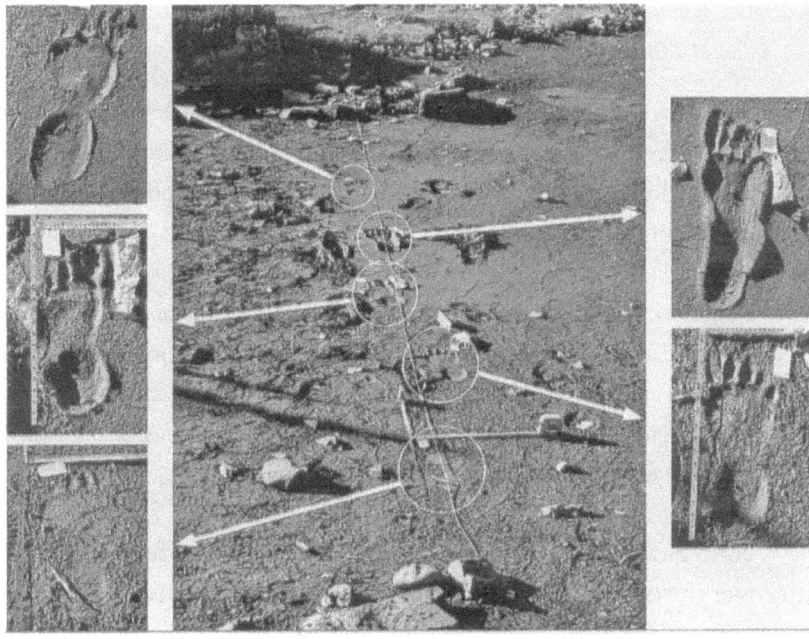

SAYWARD TRACKS

The deer hunter and his companion had stopped on a logging road to view a drying pond which appeared to have some tracks in the mud. These amateurs were able to take good photographs before the tracks were cast. They placed a rope to show the alignment of the tracks and tape measures for the length and breadth of the imprints. They had received advice to build a dam around the imprint to improve the depth of the cast. John described it as "an excellent job of documentation." With respect to a print which he considered as "borderline castable" because it was so shallow, he commented: "To their credit, they were able to cast it. It is a pretty nice cast." He always liked to give credit when credit was deserved. John went on to describe how he was making high-quality molds from these, and other, good first-generation track casts.

Last conference

The International Bigfoot Conference held at Kennewick, Washington, on September 3, 2017, was the last conference John attended. In a video interview by Rictor Riolo, he was asked why he was at the conference. His reply was twofold:

> *For people like you, who are looking for informed scientific comment, which I think I can provide, and almost more importantly for me, for me to be affirmed in what I am doing in that I continue to work as a scientist in sasquatch research, and I do not get that affirmation except at conferences like this. So thanks...to everyone for indulging me and acknowledging my efforts.*[28]

John gave a humorous presentation on the unfolding sasquatch discovery process and lack of acknowledgement received from the scientific community—a fitting topic for his last public presentation. In a spoof on his paper on sasquatch evidence being rejected yet again, he depicted himself reviving a proven attention-getting technique—flashing—as he opens his lab coat to reveal an array of clandestine evidence that he was surreptitiously peddling in the hallway of the conference. In another, playing with the idea of sasquatch as mistaken bear, he parodied a science publication, with a breaking news headline: "High-definition photographs of a Bigfoot wearing a bear disguise have scientists in damage control." This slide, spoofing the misidentified bear theory, drew laughter and applause from the audience.

In his presentation he shared from his latest writing. Field guides have entries for bears but not for sasquatches, so he had put together a parallel format for sasquatches. "This could have been done 15 years ago, maybe not as confidently as now," he said. In recognition of the value of field guides to wildlife professionals, he emphasized the need for such publications.

CHAPTER 12: AMATEUR INVESTIGATORS

JOHN INCLUDED HUMOUR IN HIS PRESENTATIONS

John ended his last presentation with an affirmation of the plight of amateur investigators and eyewitnesses, saying: "I understand how frustrating it is for investigators or people who have seen sasquatches not to be receiving the kind of informed comment that I think they are entitled to." The parallel was clear: he, as a scientist, was also frustrated at not receiving the kind of informed comment from relevant scientists that he felt entitled to.

Not believers

In many interviews, John made the point that he and other sasquatch investigators did not consider themselves to be "believers." Rather they were professionals and amateurs convinced that the evidence was compelling and warranted scrutiny. "No reputable scientist would simply be a believer," he asserted. His was an evidence-based conclusion, and he expected nothing less from other scientists. He saw the investigation of sightings and evidence as part of a scientist's professional responsibility.

John was pleased when amateur investigators adopted a similar position. He had the highest of praise for their collegiality and contribution to the field, which he summarized in the abstract for a 2013 conference paper submission: "Do biology professionals in British Columbia have a moral and ethical obligation to examine the evidence supporting the existence of the sasquatch in this province?"

> *It is becoming increasingly clear that eventual recognition of the North American sasquatch will be credited to citizen scientists. For over half a century, it has been citizen scientists—rather than relevant biologists and other scientists—who have recorded eyewitness reports, cast sasquatch tracks (as long ago as 1941), compiled the historical evidence dating back to*

the mid 1800s, and brought the subject to the attention of the few relevant scientists willing to consider it objectively.[29]

While some amateurs and professional scientists draw conclusions about the reality of the sasquatch based on evidence, the real believers are the individuals who have seen a sasquatch—the eyewitnesses.

CHAPTER 13

EYEWITNESSES:
"THEY KNOW WHAT THEY'VE SEEN"

The BC story of the ape-man that attracted John's curiosity in 1963 was the well-documented 1955 account of William Roe.[1] Hunting in the Mica Mountain area, Roe observed what at first he thought was a bear, but as it stood and walked to within 20 feet of where he was crouched down, he recognized it as a female creature, but not a bear. The thought came to him that if he shot it he "would possibly have a specimen of great interest to scientists the world over," but he felt "that it was a human being and I knew I would never forgive myself if I killed it." A hunter and trapper very familiar with western wildlife, he searched the site for additional evidence and found scat and evidence of bedding-down. Two years later, he swore a legal statement relating the details of this encounter.

Legal affidavits

Roe had such certainty that he wanted to put his encounter on record. Making a sworn statement to a legal authority is tantamount to perjury if a lie. But why did he wait two years before swearing the statement? According to the article in *True Magazine*,[2] it was John Green who encouraged Roe to make his experience known. This is not the only case of a sworn affidavit in Canada. In a second BC case the same year, an affidavit was sworn before a Justice of the Peace by Albert Ostman, giving account of his abduction by a sasquatch and subsequent captivity for six days by a sasquatch family near Toba Inlet in 1924. John Green was also familiar with his story.[3] Perhaps it was Green, an experienced newspaper man, who encouraged the use of the sworn documents. Such sworn statements lend credibility to these stories. The facts are declared to be true with a veracity akin to that of the court system, under the Canada Evidence Act.

Few eyewitnesses use the resources of the legal system to put their experiences on record. The most contemporary, and controversial, use of the judicial system involved the documentarian mentioned in the previous chapter who sued the Government of BC. Lawyers for the province argued that the lawsuit was frivolous and the judge dismissed the case for "no reasonable cause." Todd Standing is part of a group bringing a similar case before the courts in California. At the time of writing, the petition was dismissed on the advice of lawyers in order to better prepare the submission.[4]

Eyewitness reports

John diligently researched historical eyewitness reports dating back more than 150 years across North America. Sasquatches have been described, with great similarity and consistency, over this period. The earliest report presented in *The Discovery of the Sasquatch* was from Arkansas in 1851. Such accounts are important as historical facts and form part of the discovery process. Tracks, since 1941, have corroborated observations.

There is speculation that the number of reports of sightings is increasing, particularly as a result of people being in the woods with cameras. While the BFRO keeps a database of reports, which lists observations by country and province/state, it must be viewed as minimal. "In my experience most sightings do not become reports," said John. Frequently when he gave public presentations he would get new reports, usually from years past, convincing him that the frequency of sightings is higher than commonly imagined.

Media coverage is helping to raise awareness, however less than serious journalism can inhibit reports. Who would want to make a report knowing that it would not be taken seriously?

Cathartic experiences

John was responsive to the email and telephone calls that he received on what he described as the "cathartic experiences" of eyewitnesses. The term *catharsis* comes from the Greek word for cleansing. In psychology it refers to an emotional release to relieve unconscious conflicts. On the extreme side, a sighting might be a traumatic event involving threat to life and fear for safety. Or it might result in a sense of confusion and isolation. Confusion because of the lack of reference for what was experienced; isolation from peer interactions because of the perceived lack of acceptance of the truthful reported experience. "Eyewitnesses know what they've seen but need help trying to make sense of it," said John. He interpreted the best that science had to offer—the information that he was amassing. With increased experience and exposure, he became widely trusted as a scientist who would listen and help eyewitnesses to understand their encounter.

In 2016, *Psychology Today* published an article on "21 common reactions to trauma."[5] Many of these appear in eyewitness reports: In some cases eyewitnesses re-experience the trauma in their memory, including flashbacks and nightmares, particularly if there was aggression. Sightings may be accompanied by emotions such as fear and anxiety. Commonly there are changes to the way an eyewitness views the world: less trust and confidence in others and avoidance of talk about the event.

It is understandable that a witness would question what had been seen. People are rational and seek understanding within the framework of their own knowledge. When a sighting or experience is inconsistent with their personal knowledge and beliefs, it is natural to seek an explanation. Some have felt compelled to immediately write a description or make a drawing to record the event and then to seek authoritative confirmation—scientific proof—of what they saw. Others do so only after a passage of considerable time. For some, the sharing of their report with family, friends, or officials may have caused them to have self-doubts or to choose not to speak of it again. An encounter, because of the lack of reference for understanding, can leave an indelible mark in one's memory.

Delayed reports

In the two cases cited at the beginning of this chapter, there was an important difference. Roe told his story; Ostman kept his story quiet for 33 years. Was the difference due to the nature of the story? Would the story of a mistaken animal be more believable and easier to accept than a human held captive by a sasquatch family? Or were there other factors at play?

One eyewitness waited 47 years before coming forward.

> *I was only 8 years old, but I will remember it until the day I die. When we ran back and got our mom to look at the tracks, she looked at them, and*

went to the barn. She brought a garden rake out and obliterated all of the tracks she could find. I never knew why, except that both my parents came from superstitious hillbilly stock, and I guess it was too much for her. She made us both swear never to tell anyone else about any of it.

Why were the children sworn to secrecy? The mother had "freaked out" when she saw the sasquatch and had dragged the two kids into the house. When the father returned home, she only told him that "someone had been messing around the house." The 1500-word report contained substantial descriptive detail. "This incident," said the writer, "started my lifelong fascination with the big guy." He concluded with: "Sorry this is so long. I don't get to talk to anyone about this stuff very often."

Yet another recent report was withheld for an even longer period. The email began with: "I was wondering if you would be interested in an encounter I had with a very large, gorilla-like creature while camping with family when I was six years old...That was almost 50 years ago and I remember it quite vividly."

The longest-delayed report that John received was from a 75-year-old Manitoban who waited 58 years before telling his story. As a youth of 17 he was hunting moose out of season and without a license in 1941. He shot at what he thought was a moose in a willow thicket and killed a sasquatch. "I was kind of scared," he said, "as money was scarce (I had a hard time buying a box of shells). I just wanted to get the hell out of there as fast as I could." His account provided credible anatomical details of fingers, fingernails, feet, toes, height, chest, hair, ears, forehead, mouth, jaw, and face.

Why did he wait so long to come forward? Imagine the emotions—the guilt of hunting out of season, without a license, and then killing a sasquatch; and the fear of the consequence if discovered. How many times he must have contemplated how to come forward with his story. The event stayed with him, and he eventually did his own research. In his correspondence with John, he made reference to the picture drawn by Roe's daughter and the Patterson-Gimlin film. He knew that his encounter was important to tell, but acknowledged, "this is just my word. I have no evidence. That's why I want you to tell the National Geographic Society. They'll listen to you."

This last delayed report certainly had elements of fear and guilt. This story was recorded anonymously. That it was recorded at all is testimony to the hunter's recognition of the importance of the event and acknowledgement of John's credibility as a researcher and trust in his character.

Clearly, the taboo which exists for scientific discussion exists to a considerable extent in a social context, too. As a consequence, eyewitnesses consciously suppress, or unconsciously repress, any impulse or desire to talk about their encounter, sometimes for decades. This happens when they hold the memory but lack what they perceive as a safe environment in which to share. They have either experienced immediate rejection or humiliation as a result of sharing in the first instance or believe that they will experience such rejection or humiliation. As described by the above eyewitnesses, such memories do not disappear. From a psychological perspective, they may persist and contribute to anxiety or dysfunctional behaviour.

Anonymity

John has presented eyewitness accounts in his books, some with credit and some anonymous by request. Anonymity allows a person to participate, but at a distance, thus fulfilling a perceived social obligation while avoiding public embarrassment and protecting one's personal reputation. When the perceived negatives exceed the possible positives, there will be no report, or at best an anonymous report.

After providing a very detailed account of a middle-America encounter, one eyewitness wrote, "I am not sure I would want my name documented, as this is a small, rural community." He concluded with: "I would be happy to share anonymously or under a pseudonym." The perceived lack of acceptance was not worth the risk to his reputation. Eyewitness requests for anonymity were always respected by John.

Reliable information

Dean Kestner, from northern Pennsylvania, wrote that he "greatly appreciated any input from reliable sources like PhDs." As someone who had two "textbook" encounters with sasquatch behaviours and vocalizations, he knew what he had experienced. He claimed it was "very tiring leafing through the many 'unreliable' sources" who make claims of "conspiracy theories, supernatural fabrications and fables." John was cited along with Meldrum, Krantz, Green and "very few" others as researchers who present "factual, solid evidence," such as he wanted. He described how he found tracks which he could not identify for certain, but which had a divergent toe. "If it wasn't for Dr. Bindernagel's past observation concerning divergent toes associated with sasquatch, I wouldn't have been aware of that anomaly, which I have been fortunate enough to come across."[6]

From the point of view of eyewitnesses and investigators, it is important to have a reliable source for information. As Kestner noted, PhDs are usually considered a reliable source. The caveat, of course, is "if they have a genuine interest in sasquatch research."

Appeal to authority

In the early '90s, three men in their twenties, driving in the Bevan Road area of the Comox Valley on Vancouver Island, saw a sasquatch cross the road at 3:30 a.m. They went to the police to report their sighting and were asked whether they had been drinking or smoking something funny. As a consequence the sighting was recanted. What happened? Believing they had seen a sasquatch the young men went to the RCMP, an institution of authority, to report the incident. They had sufficient common knowledge to believe what they had seen. However, with the questioning they received the story was recanted. Was it the manner in which they were questioned? Did the beliefs held by the constable influence the decision to change their mind about the report?

What happens when an eyewitness cannot find authoritative corroboration? For some, the encounter with a sasquatch is a traumatic event which leaves deleterious effects similar to post-traumatic stress. John recorded one eyewitness as saying that the encounter "messed with my mind," and he sought professional counselling.

Psychologists and police are authority figures, as are scientists. What happens when those in authority do not believe in the existence of sasquatches? It is easy to understand that an eyewitness would be less likely to approach a second time. And equally easy to guess that they might generalize the lack of support to the broader group: psychologists/counsellors, police forces, and professional scientists. Under-reporting might, in part, stem from the mistrust of authority figures.

Interestingly, John spoke with one clinical psychologist who acknowledged that he had patients who thought they had seen a sasquatch. He himself, however, didn't accept that there was such a thing. How then could he guide a client to understand his or her experience? Would he try to convince someone that it was a bear, hoax, symbol or metaphor?

How one reacts to the sighting of a sasquatch depends on a number of variables, personal and situational. One avid backpacker, who had spent much of his life in the field along with his wife, had what he described as a "close interaction" in a third sighting of a sasquatch. This one was different: it "scared us because it was aggressive."

John always listened at length and provided a best of science interpretation of sasquatch encounters. His extensive field investigations lent credibility. It was his sensitivity, however, the same sensitivity that aided him in cross-cultural dialogue or in working with church youth, which was appreciated by those questioning what they had experienced. He was one authority who was very receptive.

Finding John

Examination of John's email communications from across Canada and the United States suggests that stories were related to him because eyewitnesses believed that he was someone who could be trusted. Their confidence was augmented because he was an authority figure, a PhD-qualified scientist who openly invited eyewitness reports, and who had a reputation of supporting eyewitnesses. One called John her "hero...almost unreplaceable in the sasquatch world." She related how she once asked a friend, "What should I do if I find a Bigfoot body?" The response: "Turn it over to someone really trusted...John Bindernagel."

Some people came to John without a personal reference. They found him after being attracted by his YouTube videos, televised documentaries, conference appearances, newspaper articles, or his books. Steve Pley, a Vancouver Island resident, had a second sasquatch "event." He had kept the first track to himself, but had done some research and learned of John's work online. While walking his dog between Cameron and Horne Lakes in central Vancouver Island, he came across a second track in a small, shaded patch of receding snow. He could not dismiss what he had found so he sent a photo of the track to John, "not really thinking I would receive a reply," he said. He went on to explain:

> *I was slightly apprehensive about initiating contact with John, I'm sure John receives contact from many, some legitimate, others...not so much. To be clear, I initiated contact with Dr. John Bindernagel the wildlife biologist to*

discuss a fact that I could not easily explain, not some form of entertainment entity. My expectation was zero; I didn't think I would hear anything, and when John responded (with interest and intrigue) I was quite taken aback. Once he confirmed that he wanted to go to the exact spot to investigate, I knew I had made the right decision to contact him.[7]

It was always important to John to investigate, to do an ecological assessment. This involves looking at the physical conditions: terrain, substrate, track direction, trackway measurements, food and water proximity, additional sign (e.g. dens, twisted trees, hair), and environmental conditions.

In an email of appreciation a week before John died, Steve wrote:

That day we walked the back side of Cameron Lake to take you to the spot where I found that print was a pivotal day for me; having you with me was 100% validation that the world we live in has many unknowns, yet to be mainstream and having someone who's dedicated their professional career, and more importantly their life, to that validation is something neither I nor those that share the same mindset can thank you enough for.[8]

It wasn't just that he received validation from John. The contact resulted in John taking an interest in Steve personally. He and Joan spent time with him, and John invited him to a presentation he was giving at Vancouver Island University. Steve continued:

When I think back to your presentation at VIU (so grateful you invited me!), see you on television, or when I look up at my bookcase and see your two books (that I read voraciously cover to cover numerous times), I smile and think to myself what a privilege it is to have met you and your wife in person and know that for a small window of time, we walked the same footsteps.

CHAPTER 13: EYEWITNESSES

John understood the eyewitness perspective, and he valued the eyewitnesses as he did the amateur investigators. In an interview in 2000, he asserted: "What keeps me going is people who have seen a sasquatch and say, 'I feel a lot better having talked with you.'"[9]

Importance of photographs

As with his own experience of finding tracks in Strathcona Park, people often come across evidence of sasquatches by accident. John's first question when they contacted him was always, "Did you happen to take a photograph?" With the photo capability of smart phones today, it is increasingly common for people to take photos of footprints, trackways, dens, and other suspected physical evidence. He emphasized the need to include something in the photo to indicate scale, something of known measurable size.

Mark Tressel, in contacting John for the first time on the day of John's death, wrote: "I would be interested in talking with you and sharing all of the information and photos I have. If you want, I could start sending them by email." This is the type of initial contact that would have had John bouncing and ready to go.

Serious investigators are using more sophisticated photographic equipment: GoPro cams, trail-cams, thermal cameras, infra-red cameras, camera-equipped drones. John had embraced cameras for evidence gathering. He installed motion-activated infra-red cameras on known game trails on Vancouver Island in the hope of getting a picture of a sasquatch. The technology worked well and recorded cougars, bears, deer, but never a sasquatch for him.

John: the eyewitness

John was himself an eyewitness. He believed he had seen a sasquatch while in Kentucky consulting on the Erickson Project, where a group of semi-habituated sasquatches were being observed. When asked about his observation, he explained: "It was standing in a hedgerow. We could see head and shoulder, and it was rocking back and forth." He cautioned that "it was very obscure" and while it was filmed, "it isn't a good film." What convinced him was the "big arm swinging backwards as it turned." He then expressed concern: "if someone sees that film and says that if John Bindernagel bases his acceptance of the sasquatch on a sighting like that then he is really out to lunch, I would probably agree." In describing this sighting on a *Sasquatch Chronicles* show[10] John emphasized the importance of understanding the backstory which accompanies all sightings. He also acknowledged this sighting when interviewed on his Russian trip.[11]

While he was confident that he had seen a sasquatch he did not talk about it much because it was a relatively poor sighting: partial body, obscured by hedgerow, with motion. He made the point that as a wildlife biologist he was convinced of the existence of sasquatch by the sign found—particularly tracks—as would be the case for other animals. "I do not base it on that [sighting], I base it on...tracks." He felt strongly that investigators should only bring forth the best of evidence.

CHAPTER 14

ABORIGINAL COMMUNITY:

"THEY GET IT"

Aboriginal knowledge is the oldest form of evidence supporting the sasquatch as an extant mammal. John documented indigenous records, stories, masks, totems, and petroglyphs. He was dismayed, however, that Aboriginal sightings were often simply dismissed as stories of "mythical supernatural beings." He asserted that descriptors such as "myth" and "supernatural" equate to "fictitious" for many unaware of the evidence.

The term: "sasquatch"

J. W. Burns, a teacher and Indian agent assigned to the Chehalis Band (now Sts'ailes First Nation), in the area of Harrison Lake and River, about 60 km (37 miles) east of Vancouver, BC, is credited with coining the term "sasquatch." It was popularized in 1929 in the first major Canadian article with national distribution.[1] Burns subsequently wrote more than 50 articles on sasquatch.[2]

According to the *Canadian Encyclopedia*, sasquatch is believed to be an Anglicization of the Salish word *Sasq'ets*, referring to a wild man or hairy man.[3] *Newsweek Special Edition's* "Bigfoot, The Science, Sightings and Search for America's Elusive Legend," claims the term is derived from the Halkomelem dialectal word Sésquac, with similar meaning.[4] While there may be some uncertainty with regard to the exact origin, the meaning remains clear. What is now known as the sasquatch was a creature historically understood by indigenous people of the Pacific Northwest. While pop culture in the United States favours the term "Bigfoot," sasquatch is commonly used in Canada. It stays true to its etymological roots and indigenous connection.

John felt strongly about using the sasquatch name, maintaining that it is more respectful of its cultural history and scientific identification as a species.

Cultural anthropologists

Aboriginal people have made important contributions to sasquatch research, not so much in the pursuit of investigation, until recently, but in the reporting of behaviour and activity. Indeed, the diary of explorer David Thompson records how he had heard stories from various Aboriginal people over the years about enormous creatures in the woods. He had always discounted them, "as these reports appeared to arise from the fondness for the marvellous so common to mankind." But after coming across tracks on January 7, 1811, "which measured fourteen inches in length by eight inches in breadth," which could be seen for "a full hundred yards from us," he was uncertain what to think: "The sight of the track of that large beast staggered me and I often thought of it, yet never could bring myself to believe such an animal existed."[5] Despite tracks which he took the time to measure, and a trackway that could be seen going into the distance, Thompson did not accept the stories he had heard Aboriginal people tell. Why?

Because many Aboriginal reports have been dismissed simply as myths, John said: "Not so fast!" He argued that Aboriginal people couch experience in language familiar to them, and it is the cultural anthropologists who have misinterpreted this. "Aboriginal people are telling us that sasquatches are real and are explaining their behaviours," he asserted, "and we need to listen to them and provide validation. They aren't asking for it, but deserve it."

In *The Discovery of the Sasquatch*, John gave examples from anthropological literature which have been largely interpreted as metaphor and symbol: the Wendigo of the Algonkian-speaking people, the Dzunuk'wa and Bukwus of the Kwakwaka'wakw people, and the Bushman of the Northern Athapaskan people. In their essay, "Misunderstandings arising from treating the sasquatch as a subject of cryptozoology,"[6] he and Jeff Meldrum discuss aboriginal or indigenous knowledge, including traditional accounts of wildmen, making the point that their origins may be rooted in real events.

While choosing his words carefully and not wanting to ascribe blame, John would argue that cultural anthropologists have made incorrect assumptions because the discipline itself evolved from a colonial perspective that had little understanding of, or regard for, Aboriginal life in Canada. He delighted in quoting cultural anthropologist Wayne Suttles (1918-2005), expert in Northwest Indian culture, who in discussing sasquatch cautioned against the easy categorization of myth to mean supernatural:

> We are inclined to place these creatures that are not part of our 'real' world into the category of 'mythical' or 'supernatural.' As Green has pointed out, most of us have done this with the sesqec [sasquatch] and we may be wrong.[7]

Suttles, who made seminal contributions to Northwest cultural anthropology, was humble enough to admit "we may be wrong." This was a "huge concession," said John. These words from an anthropologist of Suttles' stature were encouraging support for the arguments he was making for the sasquatch as an extant being, and for the integrity of science, where there are seldom such admissions.

John acknowledged that the sasquatch is indeed a subject of myth and legend, as are bears, orcas, beavers and a number of other animals, but this does not mean that they are not real.

"They get it"

John enjoyed visiting First Nations territories "because sasquatches are in the culture of Aboriginal people and they're generally more open to talking about them. When they see that I'm serious we have good discussions," he explained. By contrast, he said: "I don't find that going into logging camps or on the dock talking to commercial fishermen." It wasn't just that Aboriginal people generally spend more time in the wilderness, it was that they talked about sightings and kept them alive in their oral tradition. "Some of the strongest evidence for the sasquatch is Aboriginal legend, myths, dances, masks, and crests. These creatures sometimes line up extremely well with what has been reported by both natives and non-natives," explained John.[8]

John became an ambassador of science who made First Nations communities feel comfortable with a non-native researcher. He validated their beliefs, experiences, and history as he encouraged reports and respectfully listened. Then he would go into the field to search for evidence and assess the ecology. He always gave positive reports when he visited with Aboriginal eyewitnesses.

He paid great tribute to time spent in Aboriginal communities when he emphatically stated, "They get it." He didn't have to waste time and energy on explanations and answering oft-repeated questions based on personal bias or lack of knowledge. However, he did note a difference. A lot of the young people were now skeptical, but not the elders—they had seen it, or seen sign.

John greatly appreciated his Aboriginal contacts. In an email to John Green he reported: "My Aboriginal friend from Village Island dropped in this morning. He had three interesting reports...."[9] Field notes and letters indicate time spent with Aboriginal contacts in Ahousat, Hesquiat, Bella Bella, Rivers Inlet, Smith Inlet, Aristazabel Island, Lagoon Bay, Klemtu, and most recently Alert Bay. These contacts provided stories of sightings, interactions, footprints, and vocalizations.

John Zada, author of *In the Valleys of the Noble Beyond: In Search of the Sasquatch*, reached out to John and was invited to visit while on his way to conduct research with residents of the Great Bear Rainforest, an area of northern British Columbia with more than 30,000 miles of shoreline. He visited many of the places where John had followed leads. Zada concluded that "the attitude of some First Nations people yields far more because these people see the animals as a combination of physical being, spirit, story figure, symbol, and teacher. There is a definite takeaway, with manifold social and psychological benefits for both the individual and the community."

Zada continued with "all that many of the rest of us can seem to muster is the possibility of a physical ape-man and the impoverished, binary either/or debate about its existence."[10] John was frustrated with the constant debate and found relief in the attitude of people of First Nations.

Aboriginal testimony

Thomas Sewid, sasquatch investigator and administrator for the "Sasquatch Island Facebook" group, recalled meeting John after witnessing two sasquatches near Knight Inlet. "I had seen two sasquatch in my commercial fishing boat spotlight back in the early '90s. I went to see him and a true friendship was created."[11]

John encouraged Sewid to draw upon his First Nation heritage when discussing the possibility the creatures exist. "Tom—you know so much about potlatches, so much about your culture—you have got to look deeper and connect the dots." Sewid did begin to do just that and attended a sasquatch conference in 1997, where John encouraged him, saying: "Get up on that stage and just tell them what you've been telling me; they really do want to hear about your peoples' stories, perspectives and beliefs about sasquatches."

Sewid responded to John's prompt, and it became life changing:

> I did just as John asked me, and in doing so I saw and heard from the people in front of me that there really was a great interest in our peoples' perspectives towards the creature we have known about since the dawn of our creation. It was a blast of gas vapour on my sasquatch interest spark back then and it ignited a very passionate quest that I'm still on to this day! The bottom line is John's passion and how he was all a jitter and many times stammering away in his enthusiasm and excitement when the subject of the Big Fella was at hand—was just so very contagious![12]

For Sewid, this was just the beginning. He has gone on to become a popular presenter in the Pacific Northwest. Aboriginal people are now getting involved in ecotourism, and Sewid offers sasquatch tours and expeditions in the northern areas of the Salish Sea, providing an Aboriginal cultural

CHAPTER 14: ABORIGINAL COMMUNITY

perspective. While John was never keen on the commercialization of the sasquatch, he was a supporter of any Aboriginal attempt to educate the general public to their historical understanding of the sasquatch, the wild man and wild woman of the woods.

THOMAS SEWID VISITING WITH JOHN AT HOME – SEPTEMBER 2017
(THOMAS SEWID PHOTO)

A true bushman

Sewid, a former hunting guide, paid John a compliment when he wrote:

> I watched John glide through the bush world we so loved on many occasions. One day as we were getting out of my aluminum skiff to set trap cameras at a low tide, where I was ready to reach out and assist my older friend with my hand, to my surprise he stepped out upon the slippery rocks and with arms loaded glided across the boulders with ease and dexterity, while chattering, chattering, chattering, like a sasquatch! It was then at the peak of my hunting guide days that I realized this man may be a doctor, but he was the true bushman![13]

To be called a real bushman is high praise. John had an affinity with the natural world and had been with indigenous peoples in each of his overseas assignments. He was comfortable outdoors, regardless of the terrain or climate. Derek Randles offered similar acknowledgement of John's skill from an experience hiking in the Olympic Peninsula:

> I was hiking at a pretty good pace...I had been hiking for quite awhile when I decided to take a second and look back, knowing for sure I was probably leaving everybody behind. I turned around and there you [John] were, right on my heels!!! I was so impressed with your hiking ability.[14]

Already a septuagenarian, John was able to keep up a good pace on the Olympic Forest trails. Todd Prescott echoed a similar observation: "He [John] was in his early seventies when I hiked with him in Ohio, and it was all I could do to keep up with him—and I'm a little over half his age and in decent shape!"[15]

John valued physical fitness and in his senior years still enjoyed snowboarding, surfing, snorkelling, and swimming, in addition to hiking.

CHAPTER 14: ABORIGINAL COMMUNITY

Alert Bay vocalizations

John's last project was the investigation of sasquatch vocalizations at Alert Bay on Cormorant Island, a small island in Queen Charlotte Strait, east of Port MacNeil on Vancouver Island. He had first heard a "whoop, whoop, whoop" sound during an investigation at Comox Lake back in the late 1970s, which reminded him of the sounds of great apes in Africa. When he got an invitation to go to Alert Bay to investigate strange vocalizations at night, he had to go. Alert Bay is a small island community of less than 1,500 residents. There are two Kwakwaka'wakw reserves on the island. Wildlife is limited—there are no bears, cougars, or deer on the island.

INSTALLING A RECORDING DEVICE ON CORMORANT ISLAND (CTV VANCOUVER ISLAND)

The reports of possible sasquatch activity on Cormorant Island went back a decade. John had been visiting the island over the years to document sightings and record oral history. Recently a boy playing on a soccer field had seen a sasquatch and there were recurrent night time vocalizations that could not be explained by other wildlife known to be on the island. A woman was being awakened between midnight and 5 a.m. by screams and howls that were unrecognizable. She recorded them on her cell phone from her bedroom window. "Very much to her credit," commented John. He examined the cell phone recordings and then installed his own audio recording devices. He compared the sounds to the known mammals on the island and also with the sounds of birds using Cornell University's on-line library. They could not be attributed to any known wildlife.

HEADING INTO THE BUSH ON CORMORANT ISLAND (CTV VANCOUVER ISLAND)

John interviewed Alert Bay residents who heard the vocalization. One was convinced the calls were from a sasquatch, as he had experienced

CHAPTER 14: ABORIGINAL COMMUNITY

previous encounters. Once, a small tree, roots intact, was thrown at him and fellow clam diggers while harvesting on the beach of a neighbouring island.

> *Pulled the tree right out of the ground, the branches were still on it. I don't know anything that could literally pull a tree with roots and all...I mean, you see that little alder growing out there? You try and pull it out, you're not going to be able to do it.*[16]

JOHN WAS USUALLY ANIMATED WHEN TALKING ABOUT SASQUATCHES
(CTV VANCOUVER ISLAND)

Others claimed to have physically encountered the creature right on Cormorant Island. Footprints had been found on the island, and John went into the woods looking for further evidence.

This was the last project John was working on before he passed. He was welcomed by Aboriginal friends to be involved. In the BCTV news story about the vocalizations, John said:

> *It is more conceivable than people used to think, and that I used to think, that a sasquatch could be here and could not only be being observed once every several years, a fleeting glimpse, but which is leaving...a record of its presence.*[17]

Squatch and squatching

A few years before his death, John went on record to register a complaint against the casual shortening of the term *sasquatch*. He said that he would be remiss if he did not register his disappointment at the increasingly widespread use of the terms *squatch* and *squatching*. These he felt denigrated the sasquatch. He saw the contraction to *squatch* as showing a lack of sensitivity toward a cultural word of importance to Aboriginal people. He was similarly disappointed to hear of research by professionals and amateurs alike referred to as *squatching*, denigrating it as nothing more than a trivial or recreational activity. He recognized that there was no ill intent but advocated for greater cultural awareness, pointing out that non-Aboriginal investigators are "mere upstarts, adding only a little to the centuries-old knowledge that Aboriginal people have acquired" regarding the sasquatch. Always wanting to be sensitive and respectful in cultural situations, he suggested

> *that we make greater efforts to respect Aboriginal knowledge, even if it is embodied in myth and legend beyond our easy understanding. The fact that Aboriginal sasquatch knowledge has been misunderstood, even by cultural anthropologists, as fictional, does not excuse a careless or disrespectful approach to someone else's language on our part.*[18]

While some saw this as an overreaction in support of Aboriginal culture, those who knew John understood that he was applying the same standards and expectations about cultural sensitivity that he had learned to apply in his overseas assignments in Africa, the Middle East, and the Caribbean.

"We owe them"

In 2014, while speaking out about "squatch and squatching," John concluded by saying:

> *When the discovery of the sasquatch as an extant North American mammal is finally acknowledged, we will owe a huge debt to Aboriginal people for their willingness to explain the sasquatch to disbelieving anthropologists. We must affirm and applaud their efforts to educate us, to bring us onside, so to speak, just as we investigators seek the attention of relevant scientists within the scientific community to scrutinize the evidence we present for their considered (but still withheld) attention.*[19]

Currently, reconciliation is a big topic in Canada, to redress the wrongs to the Aboriginal community stemming from the colonial period of history. Cultural anthropology originated in the 19th century, the colonial period when European nations were expanding their empires around the world and were in contact with new indigenous people groups whom they conquered. John argued that there was a fundamental mistrust of Aboriginal culture and history, as it was not recorded in European tradition. Aboriginal reports were discredited simply because they were Aboriginal. In the spirit of reconciliation, John wanted the Aboriginal history of the sasquatch to be respectfully revisited.

CHAPTER 15

THE MEDIA:

"UNINFORMED OR MISINFORMED"

"The media" is a collective reference to the main means of mass communication, including publishing, broadcasting, and the Internet. The evolution of broadcasting, particularly television in the latter half of the twentieth century, revolutionized the transmittal of news information. It was the use of a motion picture camera, an early turn-of-the-century invention, however, that made the 1967 sighting of a sasquatch alongside Bluff Creek in northern California by Roger Patterson and Bob Gimlin so important. For the first time, there was actual evidence of sasquatch locomotion. Fast forward a few decades, and now anyone with a smart phone has the capacity to digitally record audio and video footage and still shots from anywhere at any time.

Mainstream media have a role to play as a watchdog of government, to protect the ideals of a democratic society, and safeguard the public interest. Collectively they are gatekeepers, just like the editors and conference chairs that denied John's submissions. They control the dissemination of news and shape the public view. They decide not only what gets in the newspapers or on the air, but how it is presented.

Generally, there appears to be a social taboo against sasquatch maintained by the media. Reports of sightings are often treated in a light-hearted, humorous manner. How often do we see a serious encounter reported with informed knowledge and objectivity? News reporters with little awareness of evidence in support of the existence of sasquatches perpetuate the hoax, myth, metaphor and misidentified bear explanations. In their defence, John argued that "they do so because they don't believe themselves. And they don't believe because relevant scientists have not examined the existing evidence and ventured a professional opinion."

The media have played a major role in the creation of the sasquatch/Bigfoot pop culture and legacy and continue to influence public perception of sasquatches. John yearned to have members of the media examine the evidence and be objective in their reporting.

The origin of "Bigfoot"

According to many sources, the term "Bigfoot" was popularized in 1958 when a road construction worker in California, Gerald Crew, brought a track cast into the *Humboldt Times* newspaper office in Eureka. Footprints had been seen by workers in the area, and whatever was making them was referred to as "big foot" by the men. An Associated Press release used the shortened noun "Bigfoot," introducing it to a wide audience. Joshua Blu Buhs, in recounting the history of Big Foot, credited the press with promoting Bigfoot "from local legend to international celebrity." He identified Andrew

Genzoli, a columnist with the *Humboldt Times*, as the one who "called the mysterious trackmaker Bigfoot, one word, which he thought played better in the newspapers."[1]

Not surprisingly, the Gerald Crew story had its skeptics. They were vindicated 44 years later, in 2002, when Ray Wallace, Crew's "prank-loving boss," died and his family came clean with the hoax story.[2] However, further examination of the evidence showed that the track casts produced by the Wallace family did not match the surviving copy of the Crew cast. Photographs of the two track casts can be seen at cryptomundo.com.[3] Herein is an example of how misinformation might impact the public. Which story is true? John pointed out that "fabricated sasquatch feet are characterized by squarish, crudely carved toes and—even more provocatively, a square heel." They are uniform, rigid, and rough-hewn and "lack flexibility, variability, and anatomical details which have been observed and documented in actual sasquatch tracks."[4] John readily acknowledged that fabricated feet exist as hoaxes, but argued that they could not account for all of the tracks found across the continent.

It is not unusual for news information to be presented with bias. Astute readers will understand this, but generally the lay public are not very discerning. They naively trust and expect the media to report the truth. They are gullible. John maintained that when the reporters are uninformed or mis-informed, the public can easily be misled. They don't have the time or interest in doing their own research, they want an instant, condensed version of the story.

Local news media

By the 1990s, John had become the "go-to sasquatch guy" for news media in Canada. The local videographer with CTV News in Courtenay, Gordon Kurbis, had a 10-year relationship with John, averaging about one story per year with him related to his books, sasquatch sightings, and most recently when he travelled to Alert Bay to do stories on John's research on possible sasquatch vocalizations.

Kurbis first met John while doing a story on a sasquatch report from the Cumberland Lake area in the late 1970s. He interviewed John Green and became aware of John Bindernagel as a well-informed new local resource. After returning to work in the Comox Valley in 2006, he used John in a sasquatch sighting story and began to develop a working relationship with him. "John was leery at first," he said with a smile, "but understandably so."[5] Kurbis believed in the existence of sasquatch and appreciated that John had a wealth of knowledge and was available locally.

In September of 2015, Kurbis heard about possible sasquatch vocalizations at Alert Bay through Facebook, then reached out to residents. When he learned that an expert was coming to investigate, and that it was John, he contacted him to coordinate how they might be there together. He felt quite guilty staying in a hotel at company expense while John was sleeping in the back of his own van. John paid for his own gas and meals, while Kurbis enjoyed the comfort of a hotel, hot shower and per diem allowance for meals. Kurbis took pity on John and bought his meals for him the two days they were together.

Two CTV news clips were produced, the first focused on the bizarre howls that had been heard and the second on John's presence to investigate—"Renowned Sasquatch expert searches for signs,"[6] read the headlines. John was filmed going into the woods searching for signs and providing informed scientific comment.

Kurbis emphasized how important it is for local media personnel in small communities to foster good relationships with local resource people. "It is not like for national media reporters, who can drop in and then leave," he said. "We need to establish long-term relationships because we will likely need to come back to them again on future stories." He pointed out how he had interviewed John for one story where the initial take on the photographic evidence was that it "looked real." Before the story was completed, however, it became evident that the videos were actually depicting a publicity stunt by a software company. Kurbis changed the focus to protect the integrity of John's input into the story.

John and Kurbis had a respectful long-term working relationship. "There are degrees of seriousness by the media," explained Kurbis. "Some are more rigorous than others." His insights speak to the reciprocal benefits of a good working relationship with a member of the media.

In response to John's assertion that media members are misinformed or uninformed, Kurbis agreed. "It depends on the interest of the reporter," he explained. "It depends on how much they want to investigate and learn. Mostly, it depends on what is on the news feeds—there isn't time to investigate every story."

International news media

The reporting of international news is a different matter, as illustrated by the Russian trip in 2011. News media reported that scientists invited to the conference were "95 percent sure" that Bigfoot was living in the Russian tundra and that they had "irrefutable evidence." Stories quickly spread, some implicating John. "Bigfoot researcher and biologist John Bindernagel claims his research group has found evidence that the yeti (a Russian "cousin" of the American Bigfoot) not only exists, but builds nests and shelters by twisting tree branches together."[7] John always said that nests and shelters needed to be studied, but had never made definitive statements about shelters he examined in North America. For someone not familiar with the individual or the situation, the reporting could be misleading. Who is *his research group*? He was part of a group of invited guests, but they were not *his* group, nor was the purpose research. But then similar articles contained other errors: his name was misspelled, reference was made to him authoring several books (really only two), and that he was an American scientist (he is definitely a Canadian). The fact that so many outlets contained essentially the same wording, even in the headlines, is testimony to the way that news stories are spread quickly via the Internet without any fact-checking.

The evidence warranted study, but John was not prepared to provide the conclusive statements that organizers seemed to want—a consensus statement with the "95 percent" language. Evidence was possible, but not necessarily probable. The statement, however, was given to press outlets, went viral, and drew a lot of response.

Later that year, Meldrum spoke at the Pennsylvania Bigfoot Conference and expressed his opinion that the Russian expedition was more of a publicity stunt than a serious scientific endeavour.

He expressed his dismay when "the press coverage was greater than the public and academic interactions."[8] Were the media knowingly complicit or just determined to create a story?

The group was taken to what was presented as a probable yeti habitation, replete with nest, hair, and footprints. "It looked like it was staged," said John. As a joke, Meldrum dropped a chocolate snack on the ground, and pointed to the "scat." John, his accomplice, picked it up and took a taste just to be sure it was scat.

When a reporter tried to provoke John into a response to the question of falsification, John gave a diplomatic response.

Reporter: *But you are also concerned about a little bit of falsification? Do you think there is something going on here?*
John: *We have to be concerned about hoaxed tracks in North America. I'm not really concerned here, but it is always a possibility. And one wants to err on the side of being too conservative, too cautious, rather than accepting something that may not be valid. Better to keep it on the shelf or certainly, like what we saw in that cave, indicates a need for more research. It is*

sufficiently convincing to say that scientists here really could spend their time well studying that area, and they are.[9]

As with sign found in North America, research was needed. John's use of the descriptor "sufficiently convincing" simply meant that it warranted the attention of scientists. The Government of Russia had announced the establishment of a physical centre for the study of Hominology, located in the same area, a commitment that impressed John.

The same reporter asked a question to uncover the matter of risk.

Reporter: *Has it [sasquatch] caused professional or personal difficulties for you?*
John: *No, I've just kept it quiet, basically. The difficulty is that one is ignored. My biologist colleagues do not want to talk about it, so that's problematic.*

This was a deferential response. It is true that in the early days John did keep his sasquatch interests quiet, however by 2011 his reputation was well established. Indeed, it was responsible for him getting the invitation to Russia. By this time he certainly had experienced conference presentation rejections, article rejections, cuts by editors, misquotes by reporters, manipulation by television producers, and the personal anxiety and anger that occurred as a result of such professional difficulties—but he knew better than to share anything personal with a reporter.

Letter to the editor

Most newspapers offer an opportunity for readers to write a "letter to the editor" to express issues of concern, often in response to a previous news story. In response to a letter published in the *Sheboygan Press*[10] which caught his attention, John posted a video on his website and YouTube in which he countered the assertions of a writer annoyed about a sasquatch article that had appeared on the front page of an earlier edition.[11] John used the letter as a focus for the video because it showed "a great deal of misunderstanding on the part of the writer," which he added "is not his fault." He felt the writer represented a large number of people who would be offended by a reputable newspaper discussing sasquatch research. The writer viewed this as a topic of pseudoscience and asserted that it was a "disservice to those of us who are expecting professional journalism." It was pretty strong criticism. The writer identified Bigfoot with supernatural beings, legend, superstition, and myth. John methodically went through the letter, explaining the lack of knowledge of the writer who was not privy to the observations being reported.

The original article had been about an upcoming sasquatch expedition related to a reality TV show in which the sponsor made a number of statements, some of which could be misleading. John made the point that "if the sasquatch claim is to be taken seriously, we need to be a little bit conservative and not use the extreme examples which stretch our credibility."

John felt that research knowledge was not fairly represented in the mass media. He could not fault the sponsor or amateur investigators for offering their opinions, even if very speculative, because scientists simply were not paying any attention. Throughout the video he defended the letter writing critic and chose his words carefully when referring to his colleagues, although at one point he did say "they haven't abandoned the discovery process; they were never here to begin with."

Tabloid material

The Contributed Papers Chair mentioned in Chapter 12 declared that until there is hard evidence for the existence of the sasquatch (i.e., a cadaver or bones) it will remain tabloid material. Tabloids are smaller in size than regular newspapers and have shorter stories. Tabloid journalism sensationalizes stories such as crime, celebrity gossip, astrology, and subjects of cryptozoology, among other topics, and attracts the impulse shopper at supermarket checkouts. It places a priority on controversy and shock value. For example, a Google search of "tabloids and sasquatches" turned up the following front-page headlines from *Weekly World News*, an American largely-fictional news tabloid that operated from 1979 to 2007, the time frame within which John received his "tabloid material" rejection.

> I was Bigfoot's love slave now I'm pregnant with his baby
> I had Bigfoot's baby
> Bigfoot baby found
> Bigfoot kept lumberjack as love slave
> I was Sasquatch sex slave
> First Bigfoot captured
> Second Bigfoot shot dead by cops

To be told that one's professional work would remain tabloid material, akin to the above, was an insult, but the fact that tabloids could produce such bizarre fiction under the guise of news rankled John. Such journalism lacked responsibility, exploited the sasquatch, and undermined serious science.

Producer and director agenda

John had been "burned" by some producers and directors of television broadcasts and documentaries. So much so that he became very guarded when agreeing to interviews or participation. Importantly, he tried to find out the underlying reason for his invitation and how much the responsible party knew about sasquatches. He had gained popularity in Canada for sasquatch comment, and he hated to turn down any opportunity to provide input, but there was always an element of risk. He had no control over the end result.

The manipulation of science for sensationalistic reasons irritated John. He did not appreciate other professionals playing loose with scientific evidence, adulterating it, or not taking it seriously enough. As a biologist trained in the rigour of scientific methodologies, he found the production methods for television documentaries questionable at times. It annoyed him when sasquatch research was trivialized and not seen as a subject of serious scientific endeavour.

There was also the imposition. It was time-consuming to set up for videotaping and preparing for interviewers. He had to clear his schedule and put other things on hold. For crews coming from out of town, there could be other expectations such as meals together or informal touring. Very seldom did he get any remuneration for his services.

The end-product, particularly with documentaries, was often disappointing. After educating a producer and recording lengthy video footage, the content could be reduced to a snippet to suit the producer's agenda. Quotes inserted out of context or without sufficient back story fell short of what John hoped for. This illustrates the point made earlier by Gord Kurbis, where outside media personnel drop in and then leave, having no concern for an ongoing working relationship. The old adage "once bitten, twice shy," remained in John's mindset. Yet more often than not he consented to participation, believing he could contribute informed comment that would be helpful, and believing that is what responsible scientists should be doing.

Reporting of science

Unfortunately, the economics of the past few decades have cut back substantially on media resources, resulting in fewer and fewer science reporters. At the same time, traditional media outlets have been replaced by online sources. Information, correct or otherwise, is now instantaneously disseminated through networks that span the world. We talk about news being "global" and the speed being "viral." It is quick and easy to get information online. How quick? By typing the word "sasquatch" into a search engine, one can get about 11,800,000 results in 0.52 seconds. There is a great deal of information available to those who have the time to sift—and sifting is so necessary.

When asked about DNA while on the *Bigfoot Tonight Show*,[12] John admitted he was still trying to understand the technological language. He then said that he had a local DNA analyst who was demystifying it for him. If someone with a PhD in a science field has difficulty with scientific language, how much more difficult is it for the average person?

Newspapers generally have an eighth-grade reading level. At this level reporters may distort scientific findings, particularly with softer sciences, where there is more subjectivity involved in the results. Reporters covering a story might have no background in science, nor even interest. Thus there is considerable room for error.

Scientifically minded readers have an interest and expect accuracy in reporting. John always felt most sympathetic toward this population and wanted to give them the best that science could offer. He understood how smaller newspapers could not have a designated science writer, but in larger networks with specialist reporters he expected a higher standard. Science requires accuracy and objectivity, and he expected the same from science reporters.

In their defence, John explained that "media will report the prevailing acceptable knowledge."[13] In a 2018 CTV news story about a hoax, a science writer declared: "Despite the multitude of sightings, the sasquatch remains a myth at this point."[14]

Pop culture

Since 1958, Bigfoot and sasquatch have maintained a place in pop culture, particularly in the Pacific Northwest. René Dahinden, a BC resident and the most notorious of early amateur sasquatch investigators, starred in a Kokanee beer television commercial featuring a sasquatch. The sasquatch became the official mascot for Kokanee beer, and subsequent commercials featured the Kokanee Ranger and his unsuccessful attempts to hunt and catch the Kokanee beer-stealing sasquatch.

An Internet search of items carrying the sasquatch or Bigfoot name reveals outdoor, biomedical, software, lumber, food, and auto products. Novelty products abound. One distributor lists: finger feet, T-shirt, air freshener, lunchbox, action figure, bandages, pocket journal, research kit, dress-up costume, mints, pins, buttons, mug, soap, socks, scarf, sweater, ornament, can cooler, wrapping paper, gift bag, stickers, colouring book, postcards, trading cards, playing cards, and notebook.[15]

Sasquatch/Bigfoot have also appeared in the movies: children's, family, science fiction, and horror films. There is something about the elusive, mysterious creature that catches public imagination. Do you remember the popularity of the 1987 *Harry and the Hendersons* movie and subsequent television series? It was about a Seattle family's friendly encounter with a Bigfoot. The movie even won an Oscar for Best Makeup.

John objected to the commercialization of sasquatch/Bigfoot because it made light of what to him was such a serious subject. It struck particularly close to home in 2010, when Vancouver hosted the Winter Olympic Games.

"Quatchi," a character based on the sasquatch, a "local legend," was one of the official mascots created for the Olympics. The merchandising budget for the Games was $46 million (Cdn). It was reported that "Previous Olympics have made as much as $100 million from mascot-related products."[16] Not only did the choice of mascot belittle sasquatch beyond the treatment afforded other animals, the economic scale of commercial profiteering insulted John's very serious shoestring research endeavours.

Media impact

In the *Wildman* (2016) documentary, John was asked whether he thought there was a relationship between the popularity of Bigfoot in the media and the reports that were recently coming out, to which he replied:

> *I don't know about the popularity. The mass media—I have to ignore it or I get too upset. They've got blinders on. There is all this misunderstanding. First of all, there's using the term 'Bigfoot.' It kind of trivializes the whole subject. Some of the TV programs have been entertainment-oriented, so our research is getting kind of dismissed or discounted or misrepresented. It is distressing, the popularizing.*[17]

The media have so much potential to assist in the presentation of science, but are driven by ratings and advertising income, and motivated by profit. It is in their best interest to perpetuate controversy to keep readers and listeners coming back for more. "It is turning off my scientific colleagues. It's even turning off many other investigators."

John felt that television producers often exploited the sasquatch, eyewitnesses, amateur investigators, and even the scientists for their economic benefit. In contrast, those who actually knew something about the sasquatch made nothing from it. He often said that "most people who do see a sasquatch do not profit. In fact, it is quite the opposite. They suffer ridicule to the point that they are reluctant to come forward." He was certain that sasquatch sightings were under-reported. Who would want to make a report and then be embarrassed in the media?

The media have their own agenda. For example, a few years ago a local newspaper announced an upcoming public presentation on the existence of "Big Foot" and described John as a "local Sasquatch hunter and biologist." John was never a sasquatch hunter, and his Canadian presentations were always about sasquatch, not Bigfoot. An editor deemed the words "hunter" and "Big Foot" better for the newspaper's purpose, but they made John shudder.

Social media

"For me, the Internet is a mixed bag—it's easier to report a sighting, but it is not vetted very well," said John. He knew the potential to damage genuine research.

Social media have brought sasquatch from daily newspapers and monthly magazines to the immediate presence of personal communication devices. Websites and applications allow users to create and share content, giving rise to a new form of hoaxing—photo manipulation. An article in *Smithsonian Magazine*[18] makes the point: "There are so many fake videos... The problem has grown worse with social media, where viral hoaxes, like drone footage of a supposed Bigfoot in a clearing in Idaho, can rack up millions of views." Loren Coleman feels there are a growing number of shams. "Technology has ruined the old cryptozoology," he says. One only has to do a search of "sasquatch/Bigfoot photos" on Pinterest to see the confusion— what is real?

And it is not just photographic evidence. There are anecdotal observations, testimonials and narratives that sound factual but may be nothing more than fiction. Now, having the background on John's character and credibility, how do you react to the following? A story on *bigfootevidence. blogspot*[19] related the death of "Fox," a sasquatch observed over a number of years. The storyteller described riding and hiking to the spot of the burial with John and a few others. They stood in front of a large hole where Fox was to be buried after his passing. There were "a number of sasquatch. A male and female were on each side of Fox, with juveniles peeking from behind trees. I then saw Fox had extremely swelled feet, and they looked almost like he was stricken with gout. His feet were bleeding as he lay on the ground." The guests approached slowly and brought their offering of favourite things they had purchased—chewing tobacco, peanut butter and shortbread. The writer described the event as "exciting and also frightening" and observed: "Upon viewing Fox, John became very emotional and began to cry. He had a difficult time maintaining his composure and tears were rolling down his cheeks."

Is there truth to this story? Had John participated in such an event he would have been agog with revelation: human/sasquatch interaction, end of life ritual, a body, burial practice, food preferences, language and communication, site ecology, and more. There would have been so much science to record and report. But John never spoke of, or wrote anything about, this happening. It appears that his name was used to lend credibility to a fabricated story.

Despite the ills associated with social media, John embraced videography as a tool for presenting research and educational information, and a personal website and Facebook channel as a means for dissemination of his material. While his sites did not garner the number of hits a hoax might have, they nevertheless did better than the sales of his books.

Media archives

Newspaper archives provide historical accounts of sasquatch dating back to the previous century, e.g., an animal bearing the unmistakable likeness of humanity (1851), gorilla or wild man (1870), wild man...hairy as an old bear (1889), some kind of monstrosity (1891). These references are cited here as illustrative, and not exhaustive. Discussion of these and other early reports can be found in *The Discovery of the Sasquatch* and contain many anatomical and behavioural details.[20] John did extensive research of newspaper accounts. He maintained that published historical accounts warrant serious examination for at least two reasons: for their existence as historical facts, and for their importance as part of the discovery process. He made the point that most people do not understand that "sasquatches have been consistently described—often in considerable detail—over the course of 150 years," and for the last 75 years these accounts have been corroborated by track casts.

While media archives played a role in the sasquatch discovery claim process that John championed, he recognized the difficulties associated with modern online research, primarily "fake news" and photo manipulation. Internet stories, correct or otherwise, can now exist in perpetuity, making future research more difficult.

PART THREE

THE LEGACY

"My mother always told me that as you go through life, no matter what you do, or how you do it, you leave a little footprint, and that's your legacy."

JAN BREWER

CHAPTER 16

TOWARD THE END:

THE LAST HURRAH

A science article in *The Globe and Mail* in 2011[1] reported that John felt that it was only a matter of time before he would be proved right about sasquatch; however, he wondered if he would still be around for his vindication. "I'm turning 70 this fall...time is running out. That's certainly what I think about quite a lot," he said. On the one hand, he was optimistic that scientific evidence in the form of DNA would soon be forthcoming, but then his hopes had been raised many times before, only to be disappointed in the end.

Fast forward to February 2017. In a *Sasquatch Syndicate* podcast on "Building the Citizen Scientist,"[2] John shared some personal reflections.

For the past 15 years he had been working full-time on sasquatch research and education "to the neglect of everything else." While working overseas he had been viewed as a "real biologist," he said. At home in Canada, it was different. "I still think I am a real biologist studying a North American mammal," he stated. However, because he wasn't employed as a biologist there was a diminished public credibility. When coupled with a focus on sasquatch, a species commonly believed to not exist, his professional status was even more suspect.

Nearing the end

"It is not getting out, and this is where I feel I have failed so badly," he lamented. "It" referred to the ongoing sasquatch discovery claim message that he had been constantly championing—the awareness and education.

Despite all that he had accomplished in the eyes of others, he fell short of his own expectations. John acknowledged all of the help from amateur investigators—the "evidence, observations and data" that they shared with him. "I have little of my own," he said somewhat apologetically.

In the last few months of his life, John initiated contact with several friends in the community with whom he wished to visit. A month before he passed, he went to see Neil and Mary Jean Crouch. "We were surprised to see him," said Mary Jean. "There was no focus on his health," said Neil. "Rather John asked a couple of important questions: Have I been the Christian I was called to be? Have I done what I was supposed to do?"[3]

It is not unusual for people to become more introspective near the end of life. David Kuhl, in his book *What Dying People Want*, points out that a terminal diagnosis can be accompanied by a flood of memories and heightened self-awareness. "A person who is terminally ill does not have unlimited time or energy. In the desire to optimize the meaningful time

CHAPTER 16: TOWARD THE END

remaining, they want the truth from doctors, from friends, from family members."[4] John received loving affirmation that day from old friends.

John had always kept his health condition a private matter. He had not wanted his medical issues to interfere with his work: no undue personal attention, no backing away from contact by others. But after exhausting all treatment options, and the relegation to pain management, he finally wanted to communicate—it was now time.

JOAN AND JOHN AT HOME – CHRISTMAS 2017 (HEATHER PHILIP PHOTO)

The email letter

An email letter was to be a courtesy to his friends across the continent. He wanted to be truthful with them about his situation—his impending death—and give them an opportunity to make contact if they wished. On

Sunday, January 7, 2018, at 5:55 John sent this email to those in his computer address book:

Hello Everyone,

I was once disappointed with a friend who died before I and others in our circle of friends became aware of just how imminent was his death.

After two years of cancer chemotherapy and a year of radiation treatment, those forms of treatment are no longer effective and my own terminal cancer is now restricted to pain management. I still remain interested in recording or documenting sasquatch reports and evidence (for which I can still offer one end of an email conversation in case you are interested). I am still trying to get evidence documented by integrating reports into my files. To this end, I appreciate receiving email messages, and will try to properly address them.

Looking forward to hearing from you with any reports you are willing to share, personal information, or just greetings, and am hopeful that I will be able to respond.

As the attached photograph illustrates (chickadee and laptop), I am being well looked-after by my family members and friends who have set me up with bird feeding trays, a laptop to keep working, and the best possible conditions.

Thanks for our past communications and may they continue at least briefly,
John

The letter was accompanied by a photograph. Chris had placed a hospital bed in the living room and erected two feeding platforms for birds at the corner windows. The picture was staged. John waited for a chickadee to be present. While he had earlier been able to sit and work at his laptop on the desk, he was no longer able to do so. Rather, he used his laptop in bed. He returned to the bed immediately after the photograph was taken.

BEDSIDE PHOTO ACCOMPANYING JOHN'S EMAIL TO FRIENDS

John spoke the truth about his health that day. He knew that he didn't have long. He invited continued contact regarding his work with sasquatch as he planned to work as long as he could—there was still lots to do.

New sasquatch reports

The email letter generated 46 heartfelt messages from across North America. There was even one from Africa. At first John talked about the new reports and questions that invited dialogue. Yes, there were some new reports that he would have liked to pursue.

A response from Nova Scotia was particularly tantalizing. Darryll Walsh shared about a manuscript in process, tentatively titled *Deconstructing Bigfoot*, based on statistical analysis of more than 7,000 reports. He invited dialogue and hinted at results that will cause an "uproar" in the Bigfoot community. How John would have liked to follow up on this invitation.

There was a Canadian report of a sighting which took place seven months earlier. Friends were sitting in the backyard in a rural area of Manitoba at sunset. The property has a river running through it, many manicured trails, and a grain field. The writer, who admitted to previously believing that the sasquatch was just a figment of wild imaginations, described her encounter. There, at the end of a trail, with bush on either side, appeared a dark figure. She continued:

"Holy Shit!!!! It's a Sasquatch!!!! Oh my God! I'm actually looking at one!!!!" Was racing through my mind! Sorry for my blasphemous words.

After about a minute, I said to my friends—"Don't turn your heads, just move your eyeballs down the trail…Do you see what I think I see?… What is that? A Bigfoot?"

The creature was standing at the left side of the trail close to the tree line. He…was standing directly facing us in a slightly hunched manner. Like a wrestler about to engage with his challenger. He had his arms at about 45 degrees and was swaying side to side. Like he was dancing to slow music. He then sidestepped to our left, his right into the bush.

Well then we all starting getting excited talking in disbelief!!! All the time with our eyes riveted on the spot! He then sidestepped once again. Repeating the same motions. Stepped back into the tree line and out again for a third time! Of course my cell phone was in the truck, and there was no way I was moving from my spot! He then went back into the bush and never reappeared. He was definitely a juvenile. About 5 foot to 5 1/2 foot in height. Dark hair throughout.

Crazy me immediately without fear and running on sheer adrenaline got up and walked straight down the trail to the spot. Talking softly and singing the ABC song! Like I was coaxing a feral kitten to come! I had no *fear or feeling of aggression or danger. Maybe not the wisest move on my part. Considering Momma and or Pops must of been very near Junior.*

To no avail of seeing him again I returned to the fire only to have my friends say I was absolutely nuts to of walked off into the darkness.

She concluded by saying that she felt "so blessed" to have had the opportunity to see the sasquatch, and then added, "I hope my account puts a big smile on your face and joy in your heart as it does mine!" Indeed, such reports always put a smile on John's face.

On the 18th, shortly after he passed, a lengthy email arrived from Mark Tressel, a BC forest professional, who provided details of his encounters with a sasquatch within a day's drive of John's home. This was a new contact. This was the type of report John would have salivated over. The encounter had occurred four days earlier and involved a sighting, footprints, and urine specimens.

In the past, John would have received the photographs, followed up with an interview, and made a site assessment.

When later queried as to why he had contacted John, Tressel wrote: "I became familiar with him from watching YouTube interviews and presentations that featured him. I gained the appreciation that he was well-respected by others interested in the topic of sasquatch...John came across as being a knowledgeable, trustworthy and good man, who obviously cared a lot about his work." He added that he was now reading *The Discovery of the Sasquatch*.

"The reason I contacted Dr. Bindernagel was with the hopes that my findings and experiences could somehow further his work. Also, I was also hoping to maybe get some guidance back from him on documenting evidence properly." This combination of altruism and pragmatism was not unusual in those making contact with John.

Messages of love

The responses to John's letter of January 7 surpassed any expectations that he held. He never anticipated the overwhelming outpouring of love and encouragement that people showed him. Recipients were shocked to receive the news but prompted to write for various personal reasons.

Don Keating, a conference organizer wrote:

> *It has been a very quick 20+ years since we've met. I've learned a good bit from you... I'm eternally grateful and in your debt to you for attending my Bigfoot Conferences in the past. You, Sir, were indeed a breath of fresh air at the events. I thank you. It's been a pleasure calling you a good friend, and an honour knowing you.*

Many expressed similar sentiments—appreciation for the manner in which he readily shared his knowledge, what they learned from him, and the fact that they had a friendship.

Some expressed appreciation for his work, especially the books, and videos:

- *I've been reading your book 'Discovery of the Sasquatch' and found your approach to the sciences absolutely enthralling and refreshing.*
- *I've followed your work and found your work refreshing as it actually tackles issues in the cognitive sciences, and science in general, and the forces that operate to define 'reality,' and I'm thoroughly enjoying your work.*
- *I hope you realize what an amazing man you are. The work you have put forth is fantastic, and thank God this subject has had someone like you in it.*
- *Your videos helped me speak about my research which I never do to a group.*

Some of the messages were intended for Joan:

- *I have been thinking about him every day and as tears stream down my face. I will never forget John.*
- *He listened to me and even used some of my papers in his work. I am forever grateful.*
- *There are many folks in many groups around North America and the world that are being informed about his passing. Thousands respected your husband's courage, fortitude and commitment.*
- *I hope you get the opportunity to read this message. Please don't feel the need to reply. I am so sorry for your loss. There are so many of us who loved and admired John.*

Succession

A question in the minds of many was succinctly put by one respondent to John's letter: "Most importantly, is there anyone yet stepping up to carry the torch and continue your files?"

While corporations and institutions have resources to engage in succession planning, independent scientists do not. John did not have a succession plan—no one had come forward with similar interests with whom he could work and bequeath his files. This had been a point of anguish at times over the last decade, when he considered the body of his work and what would eventually happen to it. Without a successor, and with a terminal diagnosis, his case files were no longer a priority. Rather, time was short and he was preoccupied with the production of his research and teaching videos.

John would say that no one is indispensable. As a field biologist he explained to others about what happens when an animal dies in the wild. A carcass becomes part of a complex ecosystem as decomposition proceeds. The death of an animal creates opportunities for other organisms. Similarly, with the loss of one human life there will be others to step up and fill the void—life will carry on, and the work will continue.

One individual, never having met John and having had encounters with "screams, smells and flanking" offered thanks for John's books and videos, which helped "to make sense out of it all." At 71, he wrote:

> I am lacking your qualifications, but thanks to your shared knowledge, I will pick up the flame and go forward for as long as I can. It may not be the full torch, but I'll at least have a match or two, out in the mountains of the Pacific Northwest. No doubt others, much younger, will do the same.

Chris Bindernagel is committed to seeing that his dad's work will be available to other researchers and the public. John's books are obtainable online through Amazon. His videos are available on sasquatchbiologist.org or YouTube. The files, however, are more difficult. Many times information was given in confidence, with the understanding that names would not be made public. Every effort will be made to protect the confidence that was guaranteed by John.

"I know you are never going to read this, but...."

Of the many tributes that came in response to the letter John sent out on January 7, some responded late, knowing that John would never get to read their tribute. Jason Williams, an amateur naturalist and sasquatch investigator, wrote sentiments shared by many others:

> I first met you in 2015 at Beachfoot, and I know I drove you and poor Joan crazy, always hanging around wanting to talk or show you my gear. You were so kind, and you were sincere. We both also discovered that the other enjoyed birding, which I found very exciting. I remember

CHAPTER 16: TOWARD THE END

several later conversations where we discussed the enjoyment we both derive from birding.

We spoke many times on the phone, but I do not believe you understood how excited I was to speak to you every time.

I thoroughly enjoyed your YouTube lectures. I looked forward to learning something new each time I saw you had posted a new video; I wish you had been able to make more of them.

I have the books that you autographed for me here on my desk. You made my day when I found them in my mailbox.

You have been a great inspiration to me, to seek evidence (I should say more evidence) of the sasquatch. When your scientific colleagues finally take this subject seriously, it will be to a great extent because of your work that they do.

I know you are never going to read this, but I just wanted to write you one last time. You did a lot of good in the world during your time, John. I know because I am one of the people who benefited from knowing you. Thank you.

The end

A week before he died, confined to bed, John responded to a call from a radio station asking for a telephone interview three days hence. "If I am still alive," he quipped with a parched voice, not wanting to say no. He knew the end was near, but there was perhaps one last opportunity to provide informed scientific comment. He was still alive three days later, but unable to do an interview.

The emails he received over the last 10 days were comforting. "They are keeping me going" he said on the 15th. That was part of our last conversation. When I visited on the 17th he was not able to focus. There was no longer any conversation. I was able to give him some liquid and talk to him as he drifted in and out of consciousness. He was pain-free and peaceful when I left.

John passed in the early hours of January 18, 2018, with Joan at his side.

News of his death was carried by newspapers and on television news across Canada and throughout North America through sasquatch/Bigfoot Internet sites and networks. Tributes appeared for him on many online sites. The words from Wes Germer on *sasquatchchronicles* express the thoughts of many:

> *John meant so much to so many people and he will truly be missed. I learned so much from John; he was the most humble, kind and intelligent person I have ever met and I was lucky enough to call him a friend.*

The next sentence reminds us of John's optimistic outlook on life: "So as John always said, 'It's never goodbye, Wes, just bye for now.'"[5]

CHAPTER 17

THE LEGACY:
HIS FOOTPRINT

It has been said that the depth of a man's convictions measures the breadth of his influence. This is an apt introduction to a discussion of John's legacy: he had deep convictions—religious, moral, ethical—which, when combined with the depth of his knowledge across several fields of science, and his indomitable personality, contributed to a wide influence. He won't be remembered simply as a sasquatch investigator, but as a *much beloved* sasquatch researcher, scientist, and friend.

Footprint metaphor

To a wildlife biologist, tracks are evidence of the presence of animals. Similarly, as humans we all leave a footprint where we tread. Today we speak of carbon footprints and their impact on the natural world. Carbon footprints have been described as small, like a "baby shoe," or huge, "like a sasquatch walking through snow."[1] Individual footprints can reveal distinguishing characteristics, for example: size and weight, direction, and length of presence. Footprints are an apt metaphor for John's life. He trod far and wide and left some lasting impressions.

John was described as a "breath of fresh air" by a conference organizer, as "hero of sasquatch research" by an amateur supporter, and as an "inspiration" to a newbie investigator. He was thanked for being "so kind, so upbeat" to wannabe researchers. As one stated, he made "a difference in my life by being supportive of my curiosity about and efforts to defend the natural world." This was a sentiment echoed by many. John's footprints should encourage others to follow.

Leaving a legacy

A legacy is that which is inherited or received from a predecessor. We all leave a legacy, but some leave more than others. Even before his diagnosis of cancer in 2016, John had an age-related concern for the question of legacy. He was getting older and realized his life work was incomplete and that there was a need to have a succession plan. With the diagnosis this concern became more urgent. However, no Canadian wildlife biologist had come forth with a similar research interest in sasquatch. What was to happen with his files?

For a scientist committed to the study and explanation of an unrecognized species that he fervently believed to exist, the thought of his work

CHAPTER 17: THE LEGACY

being lost was painful. Despite his successes, there was an irksome feeling of failure that reared its head at times, so much so that he questioned whether he had really done what he had been called to do. He had high energy and had never experienced burnout. He had always been optimistic, choosing to focus on the positive. He was resilient, bouncing back from rejections. But near the end, in the most reflective time of life, he weighed his accomplishments against his personal expectations, wrestled with the balance, and wondered what his legacy would be.

He hoped to leave a positive legacy—a contribution of some worth. Much of his information, however, was kept in his head. How to pass on his information had been on his mind for several years. Now he was running out of time and no longer had the focus and perseverance for organizational tasks. He perseverated on the story, the unfolding discovery claim for the sasquatch. It was his dying wish that it be put on record.

Toward the end he said: "I want to illustrate to readers, particularly young science students, there is a place for major scientific discoveries to take place. However, it is going to take much more fortitude and perseverance than I had anticipated." He had worked tirelessly to bring forth the story of the discovery of the sasquatch and believed that the sasquatch would someday be recognized as a species of mammal. His books, videos, presentations, and now his biography, are part of the story, part of his legacy.

A stalker?

John would be pleased that his passing was noted in *The Globe and Mail*.[2] He was an avid, lifetime reader of this newspaper. The obituary headline read "Scientist Spent Decades Stalking Sasquatch." With respect to the pursuit of animals, stalking implies the tracking of prey in a stealthy manner. Definitions of stealthy include synonyms such as: cautious, secretive, covert, surreptitious action or movement. The headline conjures images of a hunter furtively following sasquatch tracks on a routine basis for decades. John would be offended by the choice of the word *stalking*. This is something that he might associate with Bigfooters who go out in the woods to "bag" a sasquatch. He, on the other hand, did not stalk. He was out in nature to investigate sighting reports and do ecological assessment.

Cryptozoologist?

The same obituary also identified John as a cryptozoologist. It is probably true that most scientists consider the sasquatch to be a cryptid, an animal whose present existence is unproven. John, however, after the examination of considerable evidence, firmly believed in the existence of the sasquatch as a current North American mammal.

Cryptozoology is the study of "hidden animals," and is often associated with the Loch Ness monster, Ogopogo, and other creatures of legend and myth. It is considered to be pseudoscience in part because there are no established scientific methods within the discipline. John did not self-identify as a cryptozoologist. Once again, the media unwittingly conspired against him. In fact, John co-authored a published article in 2012 titled "Misunderstandings arising from treating the sasquatch as a subject of cryptozoology," in which it was suggested that delegating the sasquatch to the category of cryptozoology had more to do with lack of awareness of the evidence for its existence than the absence of such evidence.

Despite his position on sasquatch and cryptozoology, John enjoyed a good relationship with cryptozoologists. Loren Coleman, one of the world's leading crypotzoologists, wrote: "Bindernagel's heart was always in the right place," and "his legacy will be profound."[3] Dr. Paul LeBlond, another cryptozoologist, admiringly stated that John did not suffer from "academic correctness."[4] Indeed, he abandoned the strictures of academia for the freedom of the wild, to be able to follow his passion uncensored. He was admired for doing what many might like to do but lack the courage. He followed his call and left not just a set of footprints, but rather a distinguished trackway that extended over a quarter of a century. He is remembered for his unorthodoxy and sasquatch advocacy.

Wildlife biologist

John was a wildlife biologist at heart and in practice. He did briefly identify himself as a "wildlife ecologist" after returning from his time in the Serengeti, a title that focuses on how human actions affect other living things and their environment, but he was most comfortable with the wildlife biologist identification.

While he graduated from a doctoral program of zoology/veterinary science, he did not self-identify as a zoologist, nor as a parasitologist, recognizing his area of specialization. His undergrad degree in wildlife management was the basis of his practice. The fact that he had earned a PhD in a field of science, however, certainly contributed to his credibility as a wildlife biologist.

A guru

Todd Prescott had 10 years of email and telephone contact with John. He described him as "the nicest man I ever met." Even though John had a PhD, he was able to "converse with us 'common folk' without a hint of arrogance or disdain," wrote Prescott.

> I knew that John was a long-time sasquatch researcher before I began corresponding with him and ultimately meeting him, but I didn't realize just how far back it was when his interest began. His deep-seated interest

became apparent when I had a chance to peruse John Green's extensive files. Reading the correspondence between Green and Bindernagel was like stepping back in time and reliving the meeting of two gurus.[5]

Guru implies a personal and spiritual relationship with a teacher. Testimony of friends and associates speaks to John's role as a guru, although he would not have thought himself as such. Yet he was certainly a teacher, mentor and intellectual guide in matters pertaining to sasquatch. His reverence for nature was evident to those who spent time outdoors with him.

Prescott said, "His energy was infectious and each of his talks always had something new to offer. And that's one other special aspect of John; he was always learning—always searching for answers." He was a sage and a student.

A legend

In an introduction to an interview with John on *After Hours with Rictor*, Rictor Riolo stated: "There are very few living legends in the world of Bigfooting, and Dr. John Bindernagel is one of them."[6] He achieved a recognition accorded to few others.

On the Olympic Project website, he was given "legend" status with some lofty company:

From Bigfoot legends like Bob Gimlin, Ron Morehead and John Bindernagel to the latest names in cutting edge techniques and technology like Dr. Jeff Meldrum and Paul Graves, every guest speaker that attends an Olympic Project expedition will offer a wealth of knowledge.[7]

To be a legend implies a certain prominence, a celebrity status. People liked John and remembered him. He stood out as a scientist because he often stood alone, committed to a path that few were willing to take. He was not deterred by risk to his reputation. In fact, it was his reputation that garnered him legendary status. He was the first wildlife biologist to be acknowledged with such regard—making him a hard act to follow. He always wanted to encourage younger scientists, not to follow him, but to follow their love for science, for discovery. Yes, he was a celebrity at conferences, where, to his wife's astonishment, "people lined up to talk with him." In the words of Jason Stroming, an admirer, who never had a chance to meet John in person:

> *Personally, I always loved Dr. Bindernagel's swagger... a scientist who was not afraid to lend his voice and opinion to the study of Bigfoot, and those who might ridicule him be damned. He didn't seem to care about his reputation, only that the mainstream scientific community get on board and start looking at Bigfoot evidence in an objective manner, rather than dismissing it out of hand.*[8]

Perhaps the swagger and the abandon contributed to his notoriety, but it was his dedication to sasquatch research and the eyewitnesses and amateur investigators over the long haul that elevated him to legend status.

Honourary Member

John was an Honourary Member of the Northern Sasquatch Research Society in the northeastern United States. Bill Brann, founder and director, had a 30-year association with John after an introduction by their mutual friend, John Green. Brann's group has accumulated more than 200 sasquatch reports. "We had many conversations over the years. I often played a taped interview I did with a witness and shared that with Dr. John," wrote Brann. The enthusiasm that John was noted for came across even in the telephone conversations. Brann went on to say that "his enthusiasm would come across over the line: YAH! YAH! OH YAH!" [9]

John had visited with Brann and appeared on *Monster Quest* with him. Brann had done the original investigations in 1976 on the Whitehall, N.Y. encounters, and John was particularly interested in these sightings. A police officer had seen a giant seven-to-eight-feet-tall creature, which was subsequently seen by others over the next few days. Brann valued John's collaboration and described him as:

> *a true professional. Soft spoken, easy going, who made time for you even during his illness. A true gentleman to the end. The last time I spoke to John his voice was very weak. He spoke of managing the pain. John and I would often talk for an hour or more, this time was only 15 minutes....Two days later I learned of John's passing.*

The most important sasquatch article ever written

John wrote "Sasquatches in our woods," an article that appeared in the Summer 2000 issue of *Beautiful British Columbia Magazine*,[10] accompanied by a painting of a sasquatch by internationally recognized wildlife artist, Robert Bateman. "This was beyond doubt the most important article on the sasquatch ever written," asserts Christopher Murphy. "It was beautifully illustrated and I nearly fell off my chair when I saw that he had arranged to get Robert Bateman to create a painting of a sasquatch."[11]

Bateman created paintings of a sasquatch and a yeti. The "Sasquatch" original belongs to an American; the yeti painting belongs to The Sir Edmund Hillary Foundation of Canada, in Toronto. Bateman's art and John's treatment of the subject "definitely moved us up considerably in getting general scientific recognition of the sasquatch," wrote Murphy.

CHAPTER 17: THE LEGACY

While information on the magazine's circulation in 2000 was not available, today this magazine has a worldwide circulation of 300,000.

"SASQUATCH" (© ROBERT BATEMAN) HAS BECOME A POPULAR IMAGE FOR CRYPTOZOOLOGICAL CONFERENCES

Bateman is a serious artist who does not paint mermaids and unicorns. His artist's description begins:

> *The moss-draped branches barely part as a large, dark shadow moves swiftly through northwestern forest. It is a vertical shape, over six feet tall; its long legs rule out a bear. It must be a primate, but is it ape or human?*[12]

Limited-edition prints of "Sasquatch" by Robert Bateman have been popular, and the image has been used for various cryptozoological conferences. After almost two decades, to many this is still the most important magazine article on sasquatch ever published.

Expert reputation

Increasingly news media were attracted to John for informed scientific comment. He enjoyed the informal recognition of "expert," based on his science credentials and the accumulation of witness accounts dating back 150 years. However, most people in Courtenay, his place of residence for 40 years, had no idea how far his expert reputation stretched. Neil and Mary Jean Crouch related how they bumped into John and Joan on the street in downtown Courtenay in 2009 while they were home on leave from work in Egypt. Two weeks later, they were back in Cairo, turned on the television, and saw John being interviewed on the banks of the Puntledge River on Al Jazeerah, the state-funded broadcast from Qatar. This was one of those small-world moments for them. John's work with sasquatch was noteworthy in Arab countries half a world away.

Social media have brought the world community closer together. John's YouTube videos and video appearances from interviews, conferences, and documentaries can be seen around the world. He is currently featured on more YouTube videos than any other sasquatch investigator. His work on documenting the sasquatch discovery process is unmatched and will enjoy the test of time.

It was his expert status that got him invited to Russia in 2011 and to the many American Bigfoot conferences where he spoke. Yet it was his down-to-earth manner that endeared him to people.

Respect of his peers

Respect of one's peers is a measure of professional success. In reflecting on John's passing, Murphy wrote:

CHAPTER 17: THE LEGACY

John was a major pioneer in sasquatch studies. His passion for this subject surpasses all researchers I have known. Many of us have written about the sasquatch, but John took everything to the next level—the sasquatch was real; not just probably real; and all that stated on the basis of first-hand experience. What John accomplished has become one of the corner stones in the field of sasquatch studies. We celebrate his great work."[3]

CHRIS MURPHY AND JOHN – BELLINGHAM 2005 (CHRIS MURPHY PHOTO)

In 2016 John was invited to present at the 2016 International Bigfoot Conference (IBF) at Kennwick, Wash., but could not make the commitment due to his cancer treatment schedule. In 2017 he was able to make the trip, and the IBF surprised and honoured him with a Lifetime Achievement Award in recognition of the longevity and significance of his research and advocacy for sasquatches. Russell Acord, fellow researcher and conference organizer, described John's "unwavering and contagious enthusiasm." To be so honoured by his peers places him in the ranks of those who will be remembered long-term for their contributions. Humbled by the honour, he offered a humorous response when the mic was passed to him by Acord.

SHARING THE STAGE AND THE MOMENT: (L-R) DAVID ELLIS, DEREK RANDLES, JOHN, THOMAS SEWID, JEFF MELDRUM, RUSSELL ACORD IN BACKGROUND (LORI ACORD PHOTO)

Acord spoke to John on the phone the day before he passed and related how, despite the strong medication and the struggle to speak, "his message was crystal clear—find the proof, the scientific proof, and prove to these people that it is there." Acord assured him that the work would continue and that the Bigfoot community appreciated everything that he had contributed. "He inspired me, even on his way out," said Acord.

Wes Germer devoted an evening program of *sasquatchchronicles*[14] to acknowledge John's passing—"Remembering John Bindernagel." He invited a long list of John's peers to reflect on their experience with John, including: Bob Gimlin, Russell Acord, Derek Randles, Ron Morehead, Paul Graves, Marc Myrsell, Thomas Sewid, Rictor Riolo, Shane Corson, Todd

CHAPTER 17: THE LEGACY

and Diane Neiss. Germer, introduced the show by saying: "I think I can sum up the show tonight like this: John was an example of how to be a great human being. He always treated people with respect and was loved and admired by many."

IBF CONFERENCE RECOGNITION PLAQUE

Acord described him as softspoken and a gentleman and noted that you couldn't help but get caught up in his enthusiasm if you were in the room with him:

> ...his enthusiasm was just insane. He was just a bundle of energy and enthusiastic and loving and caring...That would be something that we could all just kind of take a page out of his playbook, in that we need to be more uplifting and encouraging and loving and enthusiastic about what we do.

Derek Randles also spoke of John's enthusiasm:

> ...if we were sharing something with him he would just get so excited almost like a kid...he never lost his enthusiasm, never lost his excitement to read to the end...the group email that he sent out to everybody, knowing that he was going to be passing...he still wanted people to send him Bigfoot stuff, and to me, that was just remarkable. I mean, just how he was so devoted to the subject.

In the words of one blogger, "This tribute show is a fitting beginning to cementing Doc's legacy." "Doc" was an endearing moniker used by some of John's followers in the field. Repeatedly, his peers acknowledged his contribution to sasquatch research and the character traits which made him stand out as a scientist and a person. From Russia, Dimitri Bayanov wrote: "I respected and liked John Bindernagel, for he was a daring researcher and a pleasant, nice man."[15]

Thomas Steenburg, a fellow researcher from BC, posted that John "was a clear and logical mind in the field of Sasquatch research at a time when insane thinking is becoming the norm. This, most of all, is the reason he will be missed by myself and others."[16]

Author

Jeff Meldrum has referred to both of John's books as "seminal," a term denoting their value to the field, particularly in influencing later developments. In writing a review of another author's book, he noted that "some omissions reveal a concerning degree of superficiality. For example, the seminal books by Dr. John Bindernagel should well have been included."[17]

CHAPTER 17: THE LEGACY

The Bigfoot Portal website notes his "landmark books...have garnered testimonials from top researchers and scientists like John Green, Dr. Jeff Meldrum, Dr. Jane Goodall, and Dr. Leila Hadj-Chikh." His arguments are described as "convincing," and books recommended as "excellent scholarly additions to any squatcher's library."[18]

While John had difficulty getting responses from "relevant" North American scientists, his scholarship did garner positive reviews from academics from tangential fields. The following are from his files:

- An Associate Professor of Psychology: "The Discovery of the Sasquatch *offers important understanding of both the strengths and weaknesses of Science as it is practised in the modern world... a book that will, I believe, take its place beside the works of Thomas Kuhn and Michael Polanyi as a lasting contribution to philosophy of science.*"
- An Arts & Sciences Dean Emeritus: "*John Bindernagel has given us a closely argued, cogent, convincing explanation why the evidence has not brought widespread acknowledgment that sasquatches are extant. In doing so he underscores how impressive that evidence actually is.*"
- A Professor of Resource and Environmental Management: "*A very engaging read, especially the combination of detailed compelling descriptions of recent encounters with something very strange and apparently inexplicable, and the rules we use for constructing knowledge using the scientific method.*"

Even before *The Discovery of the Sasquatch* was released, John felt self-conscious about the price. "I apologize in advance for the high price...it is the product of extensive, thoroughly documented, research, seven years in the making. That is a long time, and it was an expensive process."[19] In reviewing the book for the Wood Ape Conservancy, Alton Higgins addressed the $49 list price in light of some "disapproving and disparaging remarks" from others:

> *I realized as I initially flipped through the book, surveyed its contents, and began to study the arguments, that what I held represented much more than a seven year investment of time and money; it was a lifetime of scholarship and experience. The value of a book is not a function of the cost or quality of paper any more than a portrait is assessed in terms of the amount of paint it contains. Notwithstanding impertinent assessments of what The Discovery of the Sasquatch should cost or might be worth, it is not possible to compensate for the true 'cost' involved with the creation of Dr. Bindernagel's opus, and it behooves all of us to simply be grateful that we can share in the life and consider the thoughts of such a man.[20]*

Inspiration to creativity

Artists are often inspired to do something creative in response to the influence of another. John would have laughed at the thought of being identified as a muse, but in truth he was the inspiration for some artists of interest to the sasquatch/Bigfoot community.

Two very different artists mentioned the *sasquatchchronicles* program, "Remembering John Bindernagel," as a motivation to create something special in tribute to John. Louisiana artist Andrew Benoit uses the computer to generate art; British Columbia artist Alex Witcombe uses driftwood. Two artists, from two countries, on different sides of the continent, using very different mediums, both paid creative homage to a man they respected.

Benoit did not know John personally but wrote: "his boundless energy and childlike curiosity seemed to always be in endless supply." By adjusting an online image of John, providing a background and a drawing, he produced the picture on this page, which he titled "When the two grey haired neighbours meet." In his words:

CHAPTER 17: THE LEGACY

I envisioned a sasquatch secretly watching him for over 40 years, beginning when the young sasquatch had deep auburn-brown hair and the young PhD biologist had dark, curly hair. Both have grown old together in a sense, and although both have grayed and aged, there is no mistaking the inquisitive, vibrant, youthful energy in the eyes of the life long researcher. And finally, after 40 years of watching him, the elder forest giant steps out of the shadows and reveals himself."[21]

FINALLY THE TWO GREY-HAIRED NEIGHBOURS MEET

Canadian sculptor Alex Witcombe created a driftwood sasquatch at Rebecca Spit Provincial Park on Quadra Island, a smaller island off the coast of Vancouver Island. The creation is eight feet tall and an estimated 500 pounds. The sasquatch is one of many sculptures Witcombe has created along the trails. "Finding gnarly, twisty pieces is such a treat," he says. "Figuring out how to build larger structures also gets my brain charged up. Initial supporting structure, attachment points, what kind of fasteners to use, etc., really gets my creative drive wound up." The object is to get people out to enjoy nature and create an interaction with public art.

I've been very interested in Sasquatch stories and had listened to Dr. Bindernagel many times on the Sasquatch Chronicles podcasts. When I found out last year that he was just a drive away, I wanted to meet him, but it was too late. When I heard of his passing I knew I had to build a Sasquatch (as a personal tribute to him and his work).[22]

ALEX WITCOMBE AND HIS CREATION "MAYHEW"

CHAPTER 17: THE LEGACY

Rictor Riolo, a Las Vegas "comedian, author, artist, Bigfoot media personality and critical commentator on all things sasquatch," first met John at the Beachfoot Conference in 2014. He later welcomed John to his *The After Hours Webcast* show in February 2016. Riolo saw another side of John. "He was a very fun man, so I took the footcast and made it like it was a guitar for him."[23] He was unaware that John had played bass guitar during his high school days, or his youthful penchant for *Mad Magazine*. Riolo has created caricatures of many of the leading sasquatch investigators.

RIOLO'S CARICATURE OF JOHN IN GUITAR POSE (SURPRISINGLY SIMILAR TO THE IMAGE OF JOHN PLAYING THE GUITAR ON P. 26)

Author Susan Ketchen met John at a meeting of the Comox Valley Writers Society. At the time she was writing *Grows That Way*, the third novel in her series about a horse-loving teen with Turner syndrome and John was working on *Discovery of the Sasquatch*. He asked if they could get together to discuss her experience with the publishing industry. It was a serendipitous meeting. While they talked a bit about publishing the topic soon changed to sasquatches.

> I had always wanted to believe that sasquatches existed, but didn't think the science was strong enough. John gave me reason to believe, in fact many reasons. Every skeptical question I asked was answered with thoughtfulness, reason, and patience. And immediately I also had the inspiration for the consolidating thread for Grows That Way, the third novel in my series: I would include a sasquatch (or two). As it turned out, the novel also needed a wildlife biologist, and it was a pleasure to notice the emergence of a congenial character, suspiciously similar to John in his warmth and gentle tenacity. Although one subsequent review of my book applauded my use of 'magic realism,' I never intended it this way. In his quiet determined way, John had convinced me that sasquatches were possible.[24]

Ketchen also noted that this meeting was memorable for another reason too, a revelation of John's character:

> I have one other clear memory from that first meeting with John. My husband had recently had a heart attack and was convalescing, slowly, at home. When I mentioned this to John, he asked if my husband wanted company, and even though they had never met, John offered to visit and spend some time with him. He was the kindest guy.

CHAPTER 17: THE LEGACY

Ketchen's experience is typical of so many stories recounted in memory of John. His passion for sasquatch entered every conversation; his affable, gregarious personality entered into relationship. John endeared himself to Ketchen as he had done to others throughout his life as a fount of scientific knowledge and as a genuine caring person.

A *naturalist*

While John was widely recognized as a wildlife biologist, he was also a naturalist with a reverence for nature wherever he went. He'd had a macro experience with nature across several continents. He had walked not only the woods of North America, but also the great lakes of Uganda, the highlands of Tanzania, the mountain basins of Iran, the hardwood forests of the Caribbean, the hills and valleys of Nepal, the rain forests and coastal plains of Belize, and river deltas of southern Africa. He was a keen observer of the flora and fauna wherever he went. Nature was always an irresistible draw for him.

John had a partner in life who shared his interest in the natural world, and they raised their children with the same appreciation. Joan enjoyed birding as much as John, and they modelled their interests for children and grandchildren. "My dad never stopped teaching me about animals, for which I am so grateful," wrote Sarah. "As a kid, I remember going on so many nature walks at home and when we travelled." But it wasn't only the fauna; it was the flora, too. There was always the admonition to "Be kind to it, and don't disrupt its growing progress. Stop to look at it and see its details."

I remember one where it was a boardwalk over and through a nature trail. I would lay down on the boardwalk to look at the plants more closely, one of them being a Venus fly trap, and learning how it got its food. My dad even taught me about the local ferns and how they pollinate. I remember learning about mushrooms, which ones to pick and which ones not to, and doing mushroom spore prints. They were so cool.[25]

As Chris explained, "It was not so much knowledge that my folks passed on to us, but rather their keen interest in the world and the value that they placed on exploring it."

A grandparent

John set a high standard for any grandparent to follow, with the explorations he provided for his grandchildren. He designed experiments, built apparatus, and created an air of expectancy for learning and challenge when Jack and Francis would visit. There was a large Rube Goldberg-like water wheel that taught the physical properties of energy as it blew soap bubbles in the air, a course of ropes and swinging bridges in the trees around the yard perimeter to teach them about climbing, and plaster of Paris to practice how to cast animal tracks and footprints. Chris explained how Jack's interest in birds sprang out of experiences in the natural world, "and the presence of my dad to share in his interest gave Jack valuable support and guidance."

Sarah's children are younger and didn't see their grandparents as much, as her family lives in Washington state. Visits, however, always featured the outdoors. Sarah wrote:

> He'd be so proud of Savannah. She's the one that collects the worms when it rains and puts them back into the dirt, lets the lizard we found crawl on her, saves the potato bugs to look at under her magnifying glass. After we'd drop Tobias off at school, she'd walk slowly, looking for spiders, snails or anything, really, then stop and examine them, watching them as they went along. I try to always remember to tell her how much Grandpa would love what she's doing and how he liked to do it too, and how proud of her he'd be.[26]

CHAPTER 17: THE LEGACY

She reminds her children how much their grandpa loved and appreciated animals, large and small.

John's family legacy is related to love of, and experiences with, the natural world. As Chris concluded:

> I hope that my children will remember my dad's lively interest in, and appreciation for, the natural world, something that he maintained to his final days as he watched the chickadees and juncos visiting the feeder at his window.[27]

A man of faith

John felt "called" to be a scientist in the same way that clerics experience a call to their ministry. He believed that all Christians were called of God for a specific purpose according to their gifts and talents. His childhood dreams and adolescent ambition, which had been affirmed by family, friends, and teachers, were reinforced by continuing education and field experience. Wildlife biology was where he was called to serve. While he will be remembered for his foundational work in sasquatch research, he was every bit a complete naturalist, an expert in natural history. His academic credentials and experience the world over bestowed on him the rank of "expert." Stewardship, the personal moral obligation, and conservation, the collective social responsibility to care for the environment were cornerstones for him. Balance and harmony were ideals. He had strong views on the ills of today: dam projects, pipelines, oil spills, the Pacific plastic patch, pesticide use, clear-cuts, the loss of habitat—an ever-increasing list—and, of course, the endangering of birds. He lived and taught respect for God's creation.

Jane Goodall once said: "I don't spend much time being introspective, believe it or not. All I know is that I grew up not questioning God because that's how you are. God was there like the birds and the wind."[28] Similarly,

John didn't spend time on introspection. He experienced God in every aspect of the natural world. It wasn't until the end of life that he became more reflective. Even then, he never questioned God's existence or presence, but rather his obedience to the call of God.

While John thought that his work on earth was incomplete, that there was no legacy, many felt otherwise. His impact in the world of sasquatch research has been recognized and honoured, but perhaps even greater is the testimony to his Christian character. "His research with sasquatch is irrelevant compared to how he impacted people,"[29] said Pat Brandon. "He had such fine relationships with everyone."

"His eyes—you had to be honest with him." Honesty and integrity were as much a part of his makeup as love of life and curiosity. He lived a life of discovery and adventure, an attractive *joie de vivre* to others. As one friend commented, "The man himself is his legacy."

APPENDIX A

GLOSSARY

anthropology: study of human societies and cultures and their development
biology: study of living organisms, divided into many specialized fields that cover their morphology, physiology, anatomy, behaviour, origin, and distribution
cryptid: creature whose existence is suspected, but has not been confirmed
cryptozoology: search for and study of animals whose existence or survival is disputed or unsubstantiated
dissenter: person who holds opinions at variance with those previously, commonly, or officially held
dissonance: tension resulting from the combination of two disharmonious or unsuitable elements or positions
evolutionary biology: subfield of biology concerned with processes that over time produced the diversity of organisms on earth
extant: still in existence; surviving

great apes: large apes of a family closely related to human
hominid: primate of a family (*Hominidae*) that includes humans and their fossil ancestors and also (in recent systems) at least some of the great apes
hominology: study of homonids
mammalogy: branch of zoology concerned with mammals
paleontology: branch of science concerned with fossil animals and plants
paleoanthropology: branch of anthropology concerned with fossil hominids
parasitology: branch of biology or medicine concerned with the study of parasitic organisms
physical anthropology: branch of anthropology concerned with the study of human biological and physiological characteristics and their development
primate: mammal of an order that includes the lemurs, bushbabies, tarsiers, marmosets, monkeys, apes, and humans. They are distinguished by having hands, handlike feet, and forward-facing eyes, and, with the exception of humans, are typically agile tree-dwellers
primatology: branch of zoology that deals with primates
pseudoscience: beliefs and practices that masquerade as science but are incompatible with scientific methodology
wildlife biology: branch of biology concerned with the study of animals and how they interact with their ecosystems
zoology: study of the behaviour, structure, physiology, classification, and distribution of animals

APPENDIX B

DR. JOHN BINDERNAGEL RESEARCH VIDEOS*

1. **Sasquatch tracks photographed and cast near Sayward, B.C.**
   Presentation and discussion of sasquatch track evidence from Vancouver Island, British Columbia, documented in photographs and as track casts. An excellent example of documentation of sasquatch evidence by amateur investigators.

2. **Recent vocalizations attributed to sasquatches**
   Audio clips and spectograms of a series of loud nocturnal vocalizations recorded on a small BC island in the fall of 2016.

Audio recordings and spectograms from Alert Bay, BC, and Norway House, Manitoba.

3. **Sasquatch trackways as an important component of sasquatch track identification**
Examples of sasquatch trackways in snow are illustrated as a guide to identifying them and to differentiating sasquatch tracks and trackways from those of bears and humans. Attention is drawn to the consistency of the two most conspicuous features of a sasquatch trackway: the long stride length and the lack of straddle or trail width. The value of snow as a substrate for documenting mammal trackways is explored.

4. **Addressing common misinterpretations of sasquatch (Bigfoot) research**
Response to widespread public misperceptions of the sasquatch (Bigfoot) as revealed in a 2016 Wisconsin letter-to-the-editor. Discussion and illustration of how the unfolding discovery of the North American sasquatch is hindered by the absence of informed scientific comment, and its resulting treatment as almost exclusively a subject of entertainment.

5. **The discovery of the sasquatch as a prolonged discovery process**
Discussion of how the philosophy of science sheds light upon this prolonged scientific discovery and resisted discovery claim: Thomas Kuhn's insight regarding such discoveries as apt and helpful.

APPENDIX B: DR. JOHN BINDERNAGEL RESEARCH VIDEOS

6. **A professional biologist updates his professional colleagues regarding the legitimacy and necessity of sasquatch research**
   Presentation prepared for the 2017 annual conference of a society of British Columbia biology professionals, which includes over 80 illustrations. This presentation was rejected, citing sasquatch being regarded as a subject of cryptozoology. It provides an opportunity for professional biologists to hear and see what they missed: the perspective and research results of a professional colleague attempting to enlighten them regarding a controversial scientific discovery.

7. **The absence of the sasquatch from North American mammal field guides: Implications and consequences.**
   Introduction as to why a more complete field guide entry for the sasquatch is necessary for inclusion in authoritative mammal field guides. A proposed field guide entry compares the most conspicuous sasquatch physical features with those of an upright bear; common foot shapes and tracks are described and illustrated.

* *These videos are available at sasquatchbiologist.org and on YouTube.*

APPENDIX C

FILMOGRAPHY

2017	Discovering Bigfoot, Documentary
2016	Wildman: My Search for Sasquatch, Documentary
2013	Larger Than Life, Portal to the Unknown TV series
2013	History or Fabulous Legend, Portal to the Unknown TV series
2009	Monsterquest TV series documentary
2008	Cryptid Hunt, Documentary short
2007	Southern Fried Bigfoot, Documentary
2007	Operation Nightscream, Documentary
2007	Best Evidence, TV Series documentary
2007	Bigfoot's Reflection, Documentary
2006	Plus grand que nature, Dossiers mystére TV series
2006	Énigmes, Dossiers mystére TV series
2003	On the Trail of Bigfoot, World of Mysteries TV series
1997	Bigfoot/Ouija Boards, Strange But True TV series

APPENDIX D

INDEX OF NAMES

Acord, Kelly, *178*
Acord, Russell, *178, 251, 252*
Amin, Idi, *37, 44*
Anderson, Roy, *48*
Audubon, John James, *16, 21*

Bateman, Robert, *4, 248-49*
Bayanov, Dmitri, *136, 162*
Benoit, Andrew, *4, 256*
Bindernagel, Albert & Mona, *18, 19*
Bindernagel, Chris, *3, 8, 10, 48, 52, 60, 61, 71, 74, 82-83, 84, 85, 86, 90, 91, 102, 232, 238, 262, 263*
Bindernagel, Jack & Francis, *262*
Bindernagel (Keyes), Joan, *3, 4-11, 35-36, 39-44, 46-48, 54-56, 61, 63, 64, 70-71, 74, 76, 77, 83, 84, 88, 90-92, 98, 100, 102, 105, 106, 108, 157, 162, 163, 168, 177, 194, 231, 237, 238, 240, 250, 261*

Bindernagel (Lacey) Sarah, 3, 5, 6, 8, 61, 74, 79, 82-83, 84, 88, 90, 93, 98, 100, 261, 262
Blood, Don, 49, 79, 89
Blu Bluhs, Joshua, 212
Bonney, Rick, 153
Bouchard, Thomas, 134
Brandon, Pat, 85, 87, 264
Brann, Bill, 247-8
Brown, Roy, 6
Burns, J.W., 198
Byrne, Peter, 172

Carson, Rachel, 34
Carter, Nick, 21
Coleman, Loren, 137, 167, 224, 244
Collard, Mark, 146
Crew, Gerald, 212-13
Croft, Tyler, 168
Crouch, Neil & Mary Jean, 230, 250

Dahinden, Rene, 122, 222
de Vos, Anton, 33, 34
Dyck, Wendy, 91

Erickson, Adrian, 135

Fay, James 'Bobo', 163
Fossey, Dian, 64

Genzoli, Andrew, 213
Germer, Wes, 174, 240, 252-3,
Gimlin, Bob, 109, 151, 152, 162, 163, 190, 211, 246, 252,

APPENDIX D: INDEX OF NAMES

Gladwell, Malcolm, 138
Goodall, Jane, 64, 111, 112, 255, 263,
Gray, Colin, 104
Graves, Paul, 156-57, 163, 246, 252
Green, John

Hadj-Chikh, Leila, 115, 136, 255
Halpin, Marjorie, 122
Hiebert, Dennis, 102
Higgins, Alton, 255
Hill, Dave, 168, 169
Hook, Ernest, 140
Humphrey, Wayne, 168

Inness, Orrey, 160, 161

James, Karyn, 5

Keating, Don, 161-62
Keddie, Grant, 138
Kestner, Dean, 191
Ketchen, Susan, 260-61
Krantz, Grover, 88, 122, 131, 150, 176, 191
Kuhl, David, 230
Kuhn, Thomas, 255, 268
Kurbis, Gord, 4, 213-15

Lacey, Savannah & Tobias, 262
LeBlond, Paul, 244
Lombardo, Tony, 164, 162, 166

McTaggart-Cowan, Ian, 133

Manson, Johnny, 165-66
Meldrum, Jeff, 107, 113, 122, 123, 131, 135, 136, 159, 162, 167, 172, 176, 191, 199, 216, 246, 252, 254, 255
Moneymaker, Matt, 150
Monson, Gunnar, 177
Morehead, Ron, 167, 177, 246, 252
Murphy, Christopher, 4, 66, 136, 248, 250-51
Myrsell, Mark, 252

Neiss, Diane, 253
Neiss, Todd, 158-60, 177, 252

Nelson, Curt, 135
Noll, Richard, 122

Obote, Milton, 44
Ostman, Albert, 186, 188

Patterson, Roger, 151, 211
Perez, Daniel, 159, 163,
Pley, Steve, 193
Polanyi, Michael, 120-21, 131, 132, 255
Prescott, Todd, 4, 162, 204, 245-46

Randles, Derek, 106, 167, 176-77, 204, 252, 254
Reynolds, Vernon, 112
Riolo, Rictor, 4, 180, 246, 252, 259
Roe, William, 32, 185, 186, 188, 190

Sanderson, Ivan T., 32, 64, 66, 67, 96, 132
Schaller, George, 113
Sewid, Tom, 177, 202-04, 252

APPENDIX D: INDEX OF NAMES

Simmons, Lori, 152-55
Solunac, Alex, 4, 106, 117, 163, 164, 168-69
Solunac, Lesley, 163, 168
Standing, Todd, 173, 186
Steenburg, Thomas, 159, 177, 254
Strong, Kendra, 87
Sunstein, Cass, 139
Suttles, Wayne, 199-200

Thompson, David, 51, 198
Thornton, Betty, 85, 87
Tressel, Mark, 235

Wallace, Donald, 152
Wallace, Ray, 213
Walsh, Darryll, 172, 233
Williams, Jason, 238
Wilson, E.O., 138
Witcombe, Alex, 256-58

Yamarone, Tom, 163

Zada, John, 201

NOTES

PART ONE: THE MAN

1. Epigraph: John James Audubon Quotes. BrainyQuote.com, BrainyMedia Inc, 2019.

Chapter 2: Undergrad hopes: Wildlife conservation

1. Ivan T. Sanderson, "A New Look at America's Mystery Giant," *True Magazine*, March 1960, http://www.bigfootencounters.com/articles/truemag.htm.
2. Loren Coleman, "The meaning of cryptozoology: Who invented the term cryptozoology?" http:// www.lorencoleman.com/explore-cryptozoology.

Chapter 3: Uganda adventure: Game cropping

1. John A. Bindernagel, *Game cropping in Uganda: A report on an experimental project to utilize populations of wild animals for meat production in Uganda, East Africa, 1968*, CIDA Library.

Chapter 4: Graduate work: Parasitology pinnacle

1. John A. Bindernagel, *Abomasal Nematodes of Sympatric Wild Ruminants in Uganda, East Africa*. A thesis submitted in partial fulfillment of the requirements for the degree of Doctor of Philosophy (Veterinary Science and Zoology) at the University of Wisconsin, 1970.
2. "August 24, 1970," University of Wisconsin Archives and Records Management, 1970-1979, www.library.wisc.edu/protests-social-action.
3. John A. Bindernagel and Roy C. Anderson, "Distribution of the meningeal worm in white-tailed deer in Canada," *Journal of Wildlife Management*, 36 (1972):1349-1353.
4. John A. Bindernagel and Roy C. Anderson, "Newer aspects of the problem of meningeal worm (Pneumostrongylus tenuis)," *Proceedings of the Eighth Annual North America Moose Workshop*, Thunder Bay, Ontario, February 1972.
5. John A. Bindernagel, "The distribution and importance of meningeal worm (Parelaphostrongylos tenuis) in western Canada, 1973," Canadian Wildlife Service Contract No. CWS 7273-073.
6. R. Sachs, H. Frank and John A. Bindernagel, (1969). "New host records for mammomonogramus in African game animals through application of a simple method of collection," *Veterinary Record*, 84:(1969): 562-63.
7. John A. Bindernagel, "Liver fluke fasciola gigantica in African buffalo and antelopes in Uganda, East Africa," *Journal of Wildlife Diseases*, 8 (1972): 315-17.
8. BFRO Report #1399: A surveyor and trader for the Northwest Company comes across tracks of an unusually large, unidentified animal, http://www.bfro.net/GDB/show_report.asp?id=1399.
9. Chris Bindernagel, email to author, May 7, 2019.

NOTES

Chapter 6: Serengeti sojourn: Ideas entertained

1. John A. Bindernagel, *Wildlife utilization in Tanzania: The ecology of three wildlife areas in Tanzania with special reference to wildlife utilization*, Dar es Salaam: Food and Agriculture Organization of the United Nations, 1975.
2. All quotations from correspondence between John Green and John Bindernagel used by permission of John Green granted to Todd Prescott.

Chapter 7: British Columbia: Domestic choices

1. Sanderson, "A New Look at America's Mystery Giant," *True Magazine* (March 1960).
2. For a comprehensive description of historical Vancouver Island sightings see "Sasquatch in BC#7: Encounters from Vancouver Island." https://www.youtube.com/watch?v=P117aEaeMRY.

Chapter 8: Iran: Persian politics

1. John A. Bindernagel, *Assistance in wildlife conservation and management Iran: Red deer management and land-use conflicts in the Caspian Forest Zone of Iran*. Department of the Environment, Food and Agriculture Organization of the United Nations, 1978.
2. ABSM, abbreviation for "abominable snowman," was used by Ivan T. Sanderson in *Abominable Snowmen: Legend come to life*, 1961.

Chapter 9: Comox Valley: Home and abroad

1. Don Blood, letter to author, April 29, 2019.
2. John A. Bindernagel, *The status, distribution and management of important wildlife species in Trinidad and Tobago*, Port of Spain: Ministry of Agriculture, Lands and Food Production, Food and Agriculture Organization of the United Nations, 1984.
3. Chris Bindernagel, email to author, May 7, 2019.

4. Sarah (Bindernagel) Lacey, email to author, July 29, 2018.
5. Betty Thornton, interview by author, December 6, 2018.
6. Pat Brandon, interview by author, December 4, 2018.
7. Kendra Strong, letter to author, September 2019.

PART TWO: HIS PASSION

1. Epigraph: Ivan T. Sanderson, "A New Look at America's Mystery Giant," *True Magazine* (March 1960).

Chapter 10: Comox Valley: Evidence found

1. John A. Bindernagel, *North America's Great Ape: The Sasquatch* (Courtenay, BC: Beachcomber Press, 1998), ix.
2. Richard Watts, "Bigfoot believers contribute evidence," *Victoria Times-Colonist,* January 8, 1994.
3. John Bindernagel, email to John Green, August 23, 1998.
4. Derek Randles, interview with Wes Germer, *Sasquatch Chronicles,* SC EP:399 Remembering John Bindernagel, January 20, 2018.
5. John Bindernagel email to John Green, August 23, 1998.
6. John Bindernagel email to John Green, July 20, 1999.

Chapter 11: Relevant scientists: "They won't examine the evidence"

1. *Bigfoot's Reflection,* Bunbury Films, 2007 (Available on YouTube)
2. Bindernagel, *North America's Great Ape,* 1.
3. Richard Watts, "The battle to find sasquatch," *Victoria Times-Colonist,* September 27, 201https://www.timescolonist.com/entertainment/the-battle-to-find-sasquatch-1.21962.
4. John A. Bindernagel, *The Discovery of the Sasquatch: Reconciling Culture, History, and Science in the Discovery Process* (Courtenay, BC: Beachcomber Books, 2010), 214.

5. Ibid, 110.
6. Ibid, 116.
7. Ibid, xvi.
8. Bindernagel, North America's Great Ape, 153.
9. Bindernagel, *Discovery of the Sasquatch*, 232.
10. Ibid, 128.
11. Michael Polanyi, (1963). "The potential theory of adsorption," *Science* 141(3585): 1010-1013. Cited in *The Discovery of the Sasquatch*, p. 14.
12. John Green, letter to John Bindernagel, June, 1977.
13. Richard Noll, post to *cryptomundo.com*, October 12, 2005.
14. Joshua Blu Buhs, *Bigfoot: The Life and Times of a Legend* (Chicago: The University of Chicago Press, 2009), 210.
15. Jeff Meldrum, *Sasquatch: Legend Meets Science* (New York: Tom Doherty Associates, 2006), 224.
16. "Sasquatch expert to speak at U of G," University of Guelph News Release (October 7, 2003).
17. Bindernagel, *Discovery of the Sasquatch*, 9-10.
18. Ibid, 221.
19. *Bigfoot's Reflections*, Bunbury Films.
20. The name of the chairperson of the review committee and the name of the journal withheld at John's request.
21. Bindernagel, *Discovery of the Sasquatch*, 15.
22. Ibid, 208.
23. Tom Paulson, "A student of Sasquatch, Prof. Grover Krantz, dies," *Seattle Post-Intelligencer*, February 17, 20ohttps://www.seattlepi.com/news/article/A-student-of-Sasquatch-Prof-Grover-Krantz-dies-1080702.php.
24. Sean Dolan, "More than myth? ISU professor speaks about Sasquatch at USU," *Idaho State Journal*, November 9, 20ohttps://www.idahostatejournal.com/news/local/more-than-myth-isu-professor-speaks-about-sasquatch-at-usu/article_256d7d31-cbf6-521d-969a-06ec5dbed75b.html.

25. Michael Polanyi, *Personal Knowledge: Towards a Post-critical Philosophy* (Chicago: University of Chicago Press, 1958), 208.
26. Richard Watts, "Prints, dusk cries stir boffin to hunt island sasquatch," *Victoria Times-Colonist*, January 7, 1994. As cited in John A. Bindernagel, "The Sasquatch: An unwelcome and premature zoological discovery?" *Journal of Scientific Exploration*, 18(1), 55.
27. Richard Watts, "The battle to find sasquatch," *Victoria Times-Colonist*, September 27, 2012.
28. Nicholas Wade, "Researcher condemns conformity among his peers," *The New York Times* (July 23, 2009). *https://tierneylab.blogs.nytimes.com/2009/07/23/researcher-condemns-conformity-among-his-peers/?ref=instapundit*.
29. Ibid.
30. Adrian Erickson, "The world is not ready for what Bigfoot is." Interview by Craig Woolheater, *cryptomundo.com* (November 2, 2016).
31. Ibid.
32. John A. Bindernagel, "The ecology of an uncatalogued hominoid of the boreal forest (taiga) of North America and Eurasia." *The Relict Hominoid Inquiry*, 7:117-134 (2018). isu.edu/media/libraries/rhi/research-papers/BINDERNAGEL_Taiga.pdf
33. Dmitri Bayanov, *The Making of Hominology* (Surrey, BC: Hancock House, 2017), 71.
34. Ibid, 74.
35. Ibid, 77.
36. Bigfoot Tonight Show, September 16, 201
37. Malcolm Gladwell, *The tipping point: How little things can make a big difference* (New York: Back Bay Books, 2002).
38. Edward O. Wilson, *Consilience: The unity of knowledge* (New York: Alfred A. Knopf, 1998), 59.
39. Cass Sunstein, (2003). *Why societies need dissent* (Cambridge, MA: Harvard University Press, 2003), 71.
40. Bindernagel, *Discovery of the Sasquatch*, 228.

41. "Banquet speaker," *The Washington Wildlifer* (Spring 2013), 4.
42. John A. Bindernagel, "Research videos," http://sasquatchbiologist.org/videos/.
43. John A. Bindernagel. "Spoiler alert," Video #http://sasquatchbiologist.org/videos/#professional.
44. John Bindernagel, interview by Gord Kurbis, *CTV News*, September 23, 2015 (https://vancouverisland.ctvnews.ca/video?clipId=317686).
45. "From petroglyphs to audio recordings, B.C. scientist advocates for examination of sasquatch evidence," *CBC* (September 6, 2017).

Chapter 12: Amateur investigators: "We owe them"

1. Bigfoot Field Research Organization, "About the Bigfoot Field Researchers Organization (BFRO)." https://www.bfro.net/REF/aboutbfr.asp.
2. Coleen Kimmett, "John Green, 'Mr. Sasquatch,' leaves big shoes to fill," *TheTyee* (June 29, 2016), thetyee.ca/Culture/2016/06/29/John-Green-Big-Shoes-To-Fill/
3. Lori Simmons, email to author, March 3, 2019.
4. Paul Graves, email to author, June 3, 2018.
5. Ibid.
6. American Primate Society website. http://www.americanprimate.org/.
7. Daniel Perez, email to author, January 11, 2019.
8. Tony Lombardo, email to author, February 2, 2019.
9. Johnny Manson, email to author, September 5, 2018.
10. International Bigfoot Conference website. https://www.sponsormy-event.com/2016-international-bigfoot-conference-kennewick.
11. Dave Hill, email to author, February 15, 2019.
12. Steve Gray, email to author, March 10, 2019.
13. John Bindernagel, letter to John Green, July 2, 1997.
14. *Sasquatch Chronicles*, SC EP:399, "Remembering John Bindernagel," January 20, 2018.

15. Archie McPhee web site advertises: Funny gifts, toys, novelties, and weird things. https://mcphee.com/
16. Darryll Walsh, email to author, June 6, 2019.
17. *Bigfoot's Reflection*, Bunbury Films, 2007.
18. David Ashby,"Discovering Bigfoot – Idaho State University professor featured in sasquatch documentary on Netflix," *IdahoStateJournal*, January 13, 201https://www.idahostatejournal.com/outdoors/xtreme_idaho/discovering-bigfoot-idaho-state-university-professor-featured-in-sasquatch-documentary/article_c53b1cdf-4f1d-5b21-b093-4baf72bbd5fc.html.
19. "Sasquatch trackers lawsuit thrown out by B.C. Supreme Court," *CBC News*, September 5, 2018. CBC.ca.
20. *Sasquatch Chronicles Blog*, Dec. 21, 201https://sasquatchchronicles.com/john-bindernagel/.
21. Bindernagel, *Discovery of the Sasquatch*, Chapter 19: Conformity and dissent, 219.
22. The Olympic Project. http://www.olympicproject.com/.
23. Derek Randles, email to author, September 5, 2018.
24. Jeff Meldrum, *Sasquatch: Legend Meets Science*, 117.
25. Derek Randles, email to author, September 5, 2018.
26. Todd Neiss, "Operation Sea Monkey," https://www.gofundme.com/f/SEA-MONKEY.
27. Video: "Sasquatch tracks photographed and case near Sayward, BC," is available at sasquatchbiologist.org
28. Dr. John Bindernagel, interview by Rictor Riolo, "Presentation at the International Bigfoot Conference," November 1, 2017.
29. John Bindernagel, Conference paper submission (2013): "Do biology professionals in British Columbia have a moral and ethical obligation to examine the evidence supporting the existence of the sasquatch in this province?" (Unpublished paper).

Chapter 13: Eye witnesses: "They know what they've seen"

1. Ivan T. Sanderson, *True Magazine* (1960, March).
2. Ibid.
3. Sasquatch Canada, "Albert's adventure." https://www.sasquatchcanada.com/uploads/9/4/5/1/945132/alberts_adventure.pdf.
4. Tobia Wayland, "California court case to prove the existence of bigfoot put on pause," *Forteana News*, March 3, 2018.
5. Seth Gillihan, "21 Common reactions to trauma," *Psychology Today*, September 1, 201https://www.psychologytoday.com/ca/blog.
6. Dean Kestner, email to author, September 25, 2018.
7. Steve Pley, email to author, May 10, 2018.
8. Steve Pley, email to John Bindernagel, January 12, 2017.
9. Wendy Dyck, "Searching for Sasquatch," *InFocus Magazine* (March, 2000), 4-5.
10. sasquatchchronicles.com/sc-ep332-dr-john-bindernagel/ (21:4 ff.)
11. youtube.com/watch?v=8xsfDFRTXzA (2:09 ff.)

Chapter 14: Aboriginal Community: "They get it"

1. J.W. Burns, "Introducing B.C.'s hairy giants," *Maclean's Magazine*, April, 1929. https://archive.macleans.ca/article/1929/4/1/introducing-b-cs-hairy-giants.
2. Alex MacGillvray, "Shouldn't be captured: Nothing monstrous about sasquatch says their pal." http://www.bigfootencounters.com/articles/jwburns2.htm.
3. Nicki Thomas, "Sasquatch," *The Canadian Encyclopedia*. https://www.thecanadianencyclopedia.ca/en/article/sasquatch.
4. A guide to deciphering the differences between a yeti, sasquatch, bigfoot and more. *Newsweek Special Edition*. December 19, 2015.

5. David Thompson, *David Thompson's Narrative, 1784-1812*, Edited by Richard Glover, Toronto: Champlain Society, 1962. Cited in Elle Andra-Warner, *David Thompson: A Life of Adventure and Discovery* (Vancouver: Heritage House, 2010), 109.
6. John Bindernagel and Jeff Meldrum,"Misunderstandings arising from treating the Sasquatch as a subject of Cryptozoology," *The Relict Hominoid Inquiry* 2(2012), 81-102 https://www.isu.edu/media/libraries/rhi/essays/BINDERNAGEL_final.pd.
7. Bindernagel, *Discovery of the Sasquatch*, 166.
8. Dyck, (March, 2000), 5.
9. John Bindernagel, email to John Green, July 20, 1999.
10. John Zada, *In the Valleys of the Noble Beyond: In Search of the Sasquatch* (Vancouver, BC: Greystone Books, 2019), 224.
11. Thomas Sewid, post to *Sasquatch Island* Facebook site, January 18, 2018.
12. Thomas Sewid, telephone interview by author, February 20, 2018.
13. Thomas Sewid. "Tribute to Dr. John Bindernagel," email attachment to author, March 6, 2018.
14. Derek Randles, email to John Bindernagel, January 15, 2018.
15. Todd Prescott, email to author, July 3, 2018.
16. CTV News, "Very eerie: Bizarre howls spark Sasquatch hunt on remote B.C. island," (September 24, 2015). https://vancouverisland.ctvnews.ca/very-eerie-bizarre-howls-spark-sasquatch-hunt-on-remote-b-c-island-1.2578210.
17. Ibid.
18. Loren Coleman, "Bindernagel speaks out against 'squatch' & 'squatching,'" *Cryptozoonews,* May 18, 201http://www.cryptozoonews.com/no-squatch/
19. Ibid.

Chapter 15: The media: "Uninformed or misinformed"

1. Blu Buhs, *Bigfoot: The life and times of a legend*, 75.
2. "The origin of the Bigfoot legend," *Today I Found Out*, May 23, 2013. http://www.todayifoundout.com/index.php/2013/05/the-origin-of-the-bigfoot-legend/.
3. "This date in history: October 5, 1958," *Cryptomundo*. https://cryptomundo.com/cryptozoo-news/bf-oct-5-1958/.
4. Bindernagel, *Discovery of the Sasquatch*, 68-69.
5. Gord Kurbis, interview by author, June 4, 2018.
6. "Renowned Sasquatch expert searches for signs," CTV News (September 23, 2015).
7. Benjamin Radford, "Yeti 'nests' found in Russia?" *Livescience* (November 18, 2011). https://www.livescience.com/17104-yeti-nest-russia-evidence.html.
8. Dr. Jeff Meldrum's presentation at the Pennsylvania Bigfoot Conference, "The Bigfoot Lunch Club" (October 25, 2011). http://www.bigfootlunchclub.com/2011/10/dr-jeff-meldrums-presentation-at.html.
9. "Bigfoot biologist John Bindernagel interview," Big Country (YouTube, July 16, 2016).
10. "Bigfoot article unnecessary." *Sheboygan Press*, Letters (July 12, 2016). https://www.sheboyganpress.com/.
11. "Addressing common misinterpretations of sasquatch (bigfoot) research" (Video #4), Sasquatch Biologist website. http://sasquatchbiologist.org/videos/#misinterpretations.
12. "Dr. John Bindernagel," Bigfoot Tonight Show (September 16, 2012). https://player.fm/series/bigfoot-tonight-show/drjohn-bindernagel.
13. Ibid.
14. CTV News (August 28, 2012). https://www.ctvnews.ca/world/man-trying-to-create-bigfoot-sighting-killed-by-teen-drivers-1.932771/comments-7.319830.
15. Archie McPhee web site. https://mcphee.com/collections/bigfoot.

16. "Mascots for Vancouver 2010 Olympics based on native mythology," *Wikinews* (November 27, 2007), https://en.wikinews.org/wiki/Mascots_for_Vancouver_2010_Olympics_based_on_native_mythology
17. Mountain Beast Entertainment (2016). Wildman: My search for Sasquatch. https://www.imdb.com/title/tt6100478/.
18. Ben Crair, "Why do so many people still want to believe in Bigfoot?" *Smithsonian Magazine* (September, 2018). https://www.smithsonianmag.com/history/why-so-many-people-still-believe-in-bigfoot-180970045/.
19. "A witness to 'Fox's' death tells their story about the dying bigfoot," September 17, 2012. bigfootevidence.blogspot.
20. John Bindernagel, *Discovery of the Sasquatch*, Chapter 4.

PART THREE: THE LEGACY
1. Epigraph: Jan Brewer Quotes. BrainyQuote.com, BrainyMedia Inc, 201

Chapter 16: Toward the End

1. Mark Hume, "Taking Sasquatch from the tabloids to the science journals," *The Globe and Mail* (August 22, 2011). https://www.theglobeandmail.com/news/british-columbia/taking-sasquatch-from-the-tabloids-to-the-science-journals/article627223/.
2. "Building the Citizen scientist – Part one," Sasquatch Syndicate (February 1, 2017). https://podtail.com/en/podcast/sasquatch-syndicate/february-2017-dr-john-bindernagel/.
3. Neil and Mary Jean Crouch, interview by author, November 2, 2018.
4. David Kuhl, *What dying people want: Practical wisdom for the end of life* (New York: Public Affairs, 2002), 193.
5. "John A. Bindernagel, 1941-2018," *Sasquatch Chronicles Blog* (January 18, 2018). https://sasquatchchronicles.com/john-a-bindernagel-1941-2018/.

NOTES

Chapter 17: The Legacy

1. "Carbon footprint," Metamia: Analogy as a teaching tool, (December 6, 2014). http://www.metamia.com/analogize.php?q=carbon+footprint.
2. Tom Hawthorn, "Scientist John Bindernagel spent decades stalking the sasquatch," *The Globe and Mail* (February 26, 2018). https://www.theglobeandmail.com/news/world/scientist-john-bindernagel-spent-decades-stalking-sasquatch/article38124214/.
3. Loren Coleman, "Wildlife biologist and sasquatch researcher John Bindernagel dies." (January 18, 2018), http://www.cryptozoonews.com/bindernagel-obit/.
4. Paul LeBlond, email to author, October 5, 2018.
5. Todd Prescott, email to author, July 3, 2018.
6. Rictor Riolo, After Hours with Rictor: Dr. John Bindernagel (Part 1) (February 24, 2016). https://www.youtube.com/watch?v=D4Cbc3uQEPA.
7. "Guest speakers," Olympic Project. http://www.olympicproject.com/expeditions/guest-speakers/.
8. Jason Stroming, "Dr. John Bindernagel has died," *The Occult Section* (January 22, 2018). http://www.theoccultsection.com/2018/01/22/dr-john-bindernagel-has-died/.
9. Bill Brann, letter to author, July 27, 2018.
10. John Bindernagel, "Sasquatches in our woods." *Beautiful British Columbia Magazine* (Summer 2000), 28-32.
11. Christopher L. Murphy, interview by author, April 9, 2018.
12. Robert Bateman, "Sasquatch." http://www.artcountrycanada.com/bateman-robert-sasquatch.htm.
13. Christopher L. Murphy, "Dr. John Bindernagel: Reflections," email to author, April 9, 2018.
14. *Sasquatch Chronicles*, SCEP:399 "Remembering John Bindernagel," (January 20, 2018).
15. Dmitri Bayanov, email to author, October 7, 2018.

16. "John Bindernagel," Thomas Steenburg website (January 20, 2018). http://thomassteenburg.com/?s=Bindernagel.
17. Jeff Meldrum, "Book review: The Sasquatch seeker's Field manual: Using citizen science to uncover North America's most elusive creature," *The Relict Hominoid Inquiry* (2016), 5:32-34.
18. "John A. Bindernagel," *Bigfoot Portal*. http://www.thebigfootportal.com/john-a-bindernagel/.
19. Alton Higgins, "The Discovery of the Sasquatch according to Bindernagel," Woodape.org. http://woodape.org/index.php/news/news/185-sasquatchdiscovery.
20. Ibid.
21. Andrew Benoit, *Sasquatch Chronicles Blog* (April 30, 2018). For examples of Benoit's art visit https://www.facebook.com/pg/MyCreatureEncounter/photos/?ref=page_internal.
22. Alex Witcombe, email to author, November 26, 2018. For more of Witcombe's driftwood creations see https://driftedcreationsart.com/blogs/gallery.
23. Rictor Riolo email to author, May 10, 2019. For examples of Riolo's art see https://www.deviantart.com/rictor-riolo/gallery/
24. Susan Ketchen, email to author, February 19, 2018.
25. Sarah (Bindernagel) Lacey, email to author, May 15, 2019.
26. Ibid.
27. Chris Bindernagel, email to author, May 7, 2019.
28. Jane Goodall, "Brainy Quotes." https://www.brainyquote.com/quotes/jane_goodall_471143.
29. Pat Brandon, interview by author, December 4, 2018.

ABOUT THE AUTHOR

Dr. Terrance James is a retired educator who taught at elementary, secondary, and post-secondary levels. He enjoyed a year in Costa Rica with the International Schools Organization and a short-term in Gaza with a CIDA project. He has a M.Ed in Educational Administration and a PhD. in Educational Psychology. He is also a private practice rehabilitation consultant. He has authored/co-authored three books on Prader-Willi Syndrome, and six books on educational and local history topics.

TERRANCE JAMES AND JOHN BINDERNAGEL, JANUARY 2, 2018

Terrance and John Bindernagel enjoyed a 25 year friendship during which Terrance vicariously participated in John's sasquatch research and sympathetically supported his frustration with the scientific community. It was John's request that the unfolding story of the discovery of the sasquatch be told. Terrance has honoured that request with his friend's biography, *Sasquatch Discovered: The Biography of Dr. John Bindernagel*.

Terrance resides with his wife, Joan, in the Comox Valley on Vancouver Island, not far from Strathcona Park where John cast sasquatch footprints.

other cryptozoology titles from HANCOCK HOUSE PUBLISHERS

Sasquatch in BC
Chris Murphy
978-0-88839-721-8
5½ x 8½, sc, 528pp
$29.95

Raincoast Sasquatch
Robert Alley
978-0-88839-143-8
5½ x 8½, sc, 360pp
$29.95

Giants, Cannibals & Monsters
Kathy Strain
978-0-88839-650-1
8½ x 11, sc, 288pp
$39.95

Sasquatch in Alberta
Thomas Steenburg
978-0-88839-408-8
5½ x 8½, sc, 116 pp
$19.95

Bigfoot Film Controversy
Patterson/Murphy
978-0-88839-581-8
5½ x 8½, sc, 240pp
$22.95

Bigfoot Research
Dmitri Bayanov
978-0-88839-706-5
5½ x 8½ sc, 424pp
$29.95

Bigfoot Film Journal
Chris Murphy
978-0-88839-658-7
5½ x 8½ sc, 106pp
$29.95

The Asian Wildman
Jean-Paul Debenat
978-0-88839-719-5
5½ x 8½ sc, 176pp
$17.95

Sasquatch/Bigfoot
Thomas Steenburg
978-0-88839-685-3
5½ x 8½, sc,128pp
$12.95

Sasquatch: *mystery of the wild man*
Jean-Paul Debenat
978-0-88839-685-3
5½ x 8½ sc, 428 pp
$29.95

Best of Sasquatch Bigfoot
John Green
978-0-88839-546-7
8½ x 11, sc, 144pp
$19.95

Best of Sasquatch Bigfoot

Know the Sasquatch
Chris Murphy
978-0-88839-657-0
8½ x 11, sc, 64 pp
$34.95

Hoopa Project
David Paulides
978-0-88839-015-8
5½ x 8½, sc, 336pp
$24.95

Tribal Bigfoot
David Paulides
978-0-88839-021-9
5½ x 8½, sc, 480pp
$29.95

The Making of Hominology
Dmitri Bayanov
978-0-88839-011-0
5½ x 8½, sc, 152 pp
$19.95

Crypto Editions *an imprint of*
Hancock House Publishers
www.hancockhouse.com
info@hancockhouse.com
1-800-938-1114

www.ingramcontent.com/pod-product-compliance
Lightning Source LLC
Chambersburg PA
CBHW070723160426
43192CB00009B/1298